EQUIVALENCE AND DUALITY FOR MODULE CATEGORIES
(with Tilting and Cotilting for Rings)

This book provides a unified approach to many of the theories of equivalence and duality between categories of modules that have transpired over the last 45 years. In particular, during the past dozen or so years many authors (including the authors of this book) have investigated relationships between categories of modules over a pair of rings that are induced by both covariant and contravariant representable functors, in particular by tilting and cotilting theories.

By here collecting and unifying the basic results of these investigations with innovative and easily understandable proofs, the authors' aim is to provide an aid to further research in this central topic in abstract algebra and a reference for all whose research lies in this field.

Robert R. Colby is Professor Emeritus at the University of Hawaii and Independent Scholar at the University of Iowa.

Kent R. Fuller is a professor of mathematics at the University of Iowa.

CAMBRIDGE TRACTS IN MATHEMATICS

General Editors

B. BOLLOBÀS, W. FULTON, A. KATOK, F. KIRWAN,
P. SARNAK, B. SIMON

161 Equivalence and Duality for Module Categories

(with Tilting and Cotilting for Rings)

EQUIVALENCE AND DUALITY
FOR MODULE CATEGORIES

(with Tilting and Cotilting for Rings)

ROBERT R. COLBY

University of Hawaii and University of Iowa

KENT R. FULLER

University of Iowa

CAMBRIDGE
UNIVERSITY PRESS

PUBLISHED BY THE PRESS SYNDICATE OF THE UNIVERSITY OF CAMBRIDGE
The Pitt Building, Trumpington Street, Cambridge, United Kingdom

CAMBRIDGE UNIVERSITY PRESS
The Edinburgh Building, Cambridge CB2 2RU, UK
40 West 20th Street, New York, NY 10011-4211, USA
477 Williamstown Road, Port Melbourne, VIC 3207, Australia
Ruiz de Alarcón 13, 28014 Madrid, Spain
Dock House, The Waterfront, Cape Town 8001, South Africa

http://www.cambridge.org

First published 2004

Printed in the United States of America

Typeface Times Roman 10.25/13 pt.   *System* LaTeX 2_ε [TB]

A catalog record for this book is available from the British Library.

Library of Congress Cataloging in Publication Data
Equivalence and duality for model categories : with tilting and cotilting for rings / Robert
R. Colby and Kent R. Fuller.
p.   cm. – (Cambridge tracts in mathematics ; 161)
Includes bibliographical references and index.
ISBN 0-521-83821-5
1. Rings (Algebra)   2. Modules (Algebra)   3. Duality theory (Mathematics)   I. Colby,
Robert R. (Robert Ray), 1938–   II. Fuller, Kent R.   III. Series.
QA247.E66   2004
512′.4 – dc22   2003066663

ISBN 0 521 83821 5 hardback

Contents

Preface

Approximately forty-five years ago K. Morita presented the first major results on equivalences and dualities between categories of modules over a pair of rings. These results, which characterized an equivalence between the entire categories of right (or left) modules over two rings as being represented by the covariant Hom and tensor functors induced by a balanced bimodule that is a so-called progenerator on either side, and which characterized a duality between reasonably large subcategories of right and left modules over two rings as being represented by the contravariant Hom functors induced by a balanced bimodule that is an injective cogenerator on both sides, have come to be known as the Morita theorems.

Morita's theorems on equivalence are exemplified by the equivalence of the categories of right modules over a simple artinian ring and the right vector spaces over its underlying division ring. More than a dozen years later the second author, expanding on the relationship between the category of modules generated by a simple module and the vector spaces over its endomorphism ring, introduced the concept of a quasi-progenerator to characterize the equivalences between a subcategory of right modules over one ring that is closed under submodules, epimorphic images, and direct sums and the category of all right modules over another ring. In the interim, employing the notion of linear compactness, B. J. Müller had characterized the reflexive modules under Morita duality and given a one-sided characterization of the bimodules inducing these dualities. These results led to several investigations of the dual notion of a quasi-progenerator that we refer to as a quasi-duality module.

In the early 1980s the notion of tilting modules and the tilting theorem for finitely generated modules over artin algebras was introduced and polished in papers by S. Brenner and M. C. R. Butler, D. Happel and C. M. Ringel, and K. Bongartz to provide new insight and examples in the representation theory of artin algebras. This has proved particularly effective in the study of algebras of various representation types. The definition of a tilting module also applies to modules over arbitrary rings and, in this setting, induces torsion theories in the categories of modules over two rings and a pair of equivalences between the torsion and torsion-free parts of the torsion theories. These tilting modules and equivalences are among the principal topics of this book.

Near the end of the 1980s C. Menini and A. Orsatti introduced a type of module, which has come to be known as a ∗-module, that generalizes both quasi-progenerators and tilting modules by inducing an equivalence between a subcategory of modules over one ring that is closed under direct sums and epimorphic images and a subcategory of modules over a second ring that is closed under direct products and submodules.

At about the same time, the first author introduced a generalization of Morita duality and the notions of cotilting modules and a cotilting theorem for noetherian rings. This inspired several versions of cotilting modules and cotilting theorems that are further principal topics of this book.

Just as progenerators, quasi-progenerators, and tilting modules are all ∗-modules, so are Morita duality modules, quasi-duality modules, and the various flavors of cotilting modules, all dual versions of ∗-modules, a kind of module that we call costar modules. As is so often the case in mathematics, generalizations of a concept led to a better understanding of the concept. Here we provide unified proofs regarding the various types of equivalence and duality, and the modules that induce them, by approaching them via the general notions of ∗-modules and costar modules. We feel that this approach should yield improved accessibility to, and better understanding of, these concepts. En route we present much of the relatively little that is known about how properties of (the modules over) one of the rings in question are transformed to the other ring under these various equivalences and dualities. We hope that this exposition will inspire further research in this and related directions.

This book contains some of the work of many authors that has inspired and, to a small extent been inspired by, our own research. We have made an attempt to give credit where it is due, but surely we have inadvertently omitted references to some works that should have been mentioned. To the authors of these works, we offer our sincere apologies, and note that they have undoubtedly been referred to in one or more of the papers and/or books in our bibliography.

Acknowledgment

Colby wishes to express his gratitude for the hospitality of the University of Iowa, where he is an independent scholar.

1

Some Module Theoretic Observations

We begin with a chapter consisting of several general facts involving various closure properties of certain categories of modules. These results are part of the background necessary for our future chapters, and we believe that they are of interest in themselves.

Throughout this book R denotes an associative ring with identity $1 \in R$, and Mod-R and R-Mod represent the categories of right and left R-modules and homomorphisms, while mod-R and R-mod denote their subcategories of finitely generated modules.

1.1. The Kernel of $\mathrm{Ext}^1_R(V, \_)$

For any R-module V we denote the kernel of $\mathrm{Ext}^1_R(V, \_)$ by $V^{\perp}$. Closure properties of $V^{\perp}$ are related to both homological and module-theoretic properties of V.

We denote the *projective dimension* of a module M by $\mathrm{proj.dim}.M$.

Proposition 1.1.1. $V^{\perp}_R$ *is closed under factors if and only if* $\mathrm{proj.dim}.V_R \leq 1$.

Proof. If $V^{\perp}_R$ is closed under factors, $M \in \mathrm{Mod}\text{-}R$, and $E(M)$ is the injective envelope of M, then, since $E(M)/M \in V^{\perp}$, the exactness of the sequence

$$0 = \mathrm{Ext}^1_R(V, E(M)/M) \to \mathrm{Ext}^2_R(V, M) \to \mathrm{Ext}^2_R(V, E(M)) = 0$$

implies $\mathrm{proj.dim}.V_R \leq 1$. Conversely, if $\mathrm{proj.dim}.V_R \leq 1$ and $M \in V^{\perp}$ with K a submodule of M, we obtain $M/K \in V^{\perp}$ from the exactness of the sequence

$$0 = \mathrm{Ext}^1_R(V, M) \to \mathrm{Ext}^1_R(V, M/K) \to \mathrm{Ext}^2_R(V, K) = 0. \quad \blacksquare$$

Proposition 1.1.2. *If* $V_R \in \mathrm{Mod}\text{-}R$ *is finitely presented, then* $\mathrm{Ext}^1_R(V, \_)$ *commutes with direct sums, so* $V^{\perp}_R$ *is closed under direct sums.*

1

Proof. If $\{M_\alpha\}_{\alpha \in A}$ is a family of modules in Mod-R, the natural monomorphism $\phi_V : \oplus_A \operatorname{Hom}_R(V, M_\alpha) \to \operatorname{Hom}_R(V, \oplus_A M_\alpha)$ is an isomorphism whenever V is finitely generated by [1, Exercise 16.3]. Moreover, ϕ induces natural homomorphisms $\theta_M : \oplus_A \operatorname{Ext}^1_R(M, M_\alpha) \to \operatorname{Ext}^1_R(M, \oplus_A M_\alpha)$. By hypothesis there is an exact sequence $0 \to K \to P \to V \to 0$ with P, K finitely generated and P projective. We obtain the commutative diagram with exact rows

$$\oplus_A \operatorname{Hom}_R(P, M_\alpha) \to \oplus_A \operatorname{Hom}_R(K, M_\alpha) \to \oplus_A \operatorname{Ext}^1_R(V, M_\alpha) \to 0$$
$$\downarrow \phi_P \qquad\qquad \downarrow \phi_K \qquad\qquad \downarrow \theta_V$$
$$\operatorname{Hom}_R(P, \oplus_A M_\alpha) \to \operatorname{Hom}_R(K, \oplus_A M_\alpha) \to \operatorname{Ext}^1_R(V, \oplus_A M_\alpha) \to 0$$

from which the lemma follows. ∎

We note that a partial converse of this last result is found in the proof of Lemma 1.2 of [77].

Proposition 1.1.3. *If V_R is finitely generated and $V_R^\perp$ is closed under factors and direct sums, then V_R is finitely presented.*

Proof. We have proj.dim.$V_R \le 1$ by Proposition 1.1.1; thus, since V_R is finitely generated, there is an exact sequence $0 \to L \to R^n \to V \to 0$, where L is projective. Hence, there is a split monomorphism $j : L \to R^{(X)}$ for some set X. By hypothesis $E(R)^{(X)} \in V^\perp$, so the composition of j with the inclusion i of $R^{(X)}$ into $E(R)^{(X)}$ has an extension to an element $f \in \operatorname{Hom}(R^n, E(R)^{(X)})$. Then $f(R^n) \subseteq E(R)^{(F)} \subseteq E(R)^{(X)}$ for some finite subset F of X. It follows that $j(L) \subseteq R^{(F)} \subseteq R^{(X)}$; therefore, since j is split monic, L is finitely generated. ∎

1.2. Gen(V) and Finiteness

We recall (see [1]) that for any collection $\mathcal{V}$ of R-modules, Gen($\mathcal{V}$) (gen($\mathcal{V}$)) denotes the full category of R-modules that are epimorphic images of (finite) direct sums of modules isomorphic to those in $\mathcal{V}$, and we let $\operatorname{Tr}_{\mathcal{V}}(M)$ denote the *trace of $\mathcal{V}$ in M*, the unique largest submodule of M that belongs to Gen($\mathcal{V}$). If $\mathcal{V}$ consists of a single module V_R we simply write Gen(V_R), and if $S = \operatorname{End}(V_R)$, then $\operatorname{Tr}_V(M)$ is the image of the canonical mapping $\nu_M : V \otimes_S \operatorname{Hom}_R(V, M) \to M$.

In order to characterize when Gen(V_R) is closed under direct products, we employ the following notions and lemmas.

Given that $\{M_\alpha\}_{\alpha \in A}$ is a family in Mod-R, for each $N \in R$-Mod we let

$$\eta_{\Pi_A M_\alpha, N} : (\Pi_A M_\alpha) \otimes_R N \longrightarrow \Pi_A (M_\alpha \otimes_R N)$$

denote the canonical mapping to obtain a natural transformation

$$\eta_{\Pi_A M_\alpha} : ((\Pi_A M_\alpha) \otimes_R \_) \longrightarrow \Pi_A (M_\alpha \otimes_R \_).$$

Lemma 1.2.1. *Suppose that $\{M_\alpha\}_{\alpha \in A}$ is a family in Mod-R. If $_R N$ is finitely generated (finitely presented), then the canonical homomorphism*

$$\eta_{\Pi_A M_\alpha, N} : (\Pi_A M_\alpha) \otimes_R N \longrightarrow \Pi_A (M_\alpha \otimes_R N)$$

is an epimorphism (isomorphism).

Proof. If $\{x_1, \ldots, x_n\}$ generate $_R N$, then any element of $M_\alpha \otimes_R N$ can be written in the form $\Sigma_i \, m_{\alpha i} \otimes x_i$.

Now assume that N is finitely presented and let $0 \to K \to P \to N \to 0$ be an exact sequence with P finitely generated and projective and K finitely generated. Then we have a commutative diagram

$$
\begin{array}{ccccccc}
(\Pi_A M_\alpha) \otimes K & \to & (\Pi_A M_\alpha) \otimes P & \to & (\Pi_A M_\alpha) \otimes N & \to & 0 \\
\eta_{\Pi_A M_\alpha, K} \downarrow & & \eta_{\Pi_A M_\alpha, P} \downarrow & & \eta_{\Pi_A M_\alpha, N} \downarrow & & \\
\Pi_A (M_\alpha \otimes K) & \to & \Pi_A (M_\alpha \otimes P) & \to & \Pi_A (M_\alpha \otimes N) & \to & 0
\end{array}
$$

with exact rows, in which $\eta_{\Pi_A M_\alpha, K}$ is epic, and $\eta_{\Pi_A M_\alpha, P}$ is easily seen to be an isomorphism by naturalness of $\eta_{\Pi_A M_\alpha}$. Hence, by the Five Lemma, $\eta_{\Pi_A M_\alpha, N}$ is an isomorphism. ∎

Identifying $R \otimes_R N = N$, we have the following result.

Lemma 1.2.2. *Let $N \in R$-Mod. Then the canonical homomorphism*

$$\eta_{R^A, N} : (R^A) \otimes_R N \longrightarrow N^A$$

is an epimorphism (isomorphism) for all sets A if and only if N is finitely generated (finitely presented).

Proof. The condition is sufficient in either case by Lemma 1.2.1. Conversely, letting $A = N$, if the diagonal element $(n)_{n \in N}$ is the image of some element $\Sigma_{i=1}^m (a_{ni})_{n \in N} \otimes_R x_i$, then, for all $n \in N$, $n = \Sigma_i a_{ni} x_i$. Thus, $_R N$ is finitely generated whenever $\eta_{R^N, N}$ is epic. Now supposing that $\eta_{R^A, N}$ is an isomorphism for all sets A, there is an exact sequence $0 \to K \to P \to N \to 0$

with P finitely generated and projective. Then both $\eta_{R^A,P}$ and $\eta_{R^A,N}$ are isomorphisms in the commutative diagram

$$
\begin{array}{ccccccc}
R^A \otimes K & \longrightarrow & R^A \otimes P & \longrightarrow & R^A \otimes N & \longrightarrow & 0 \\
\eta_{R^A,K} \downarrow & & \eta_{R^A,P} \downarrow & & \eta_{R^A,N} \downarrow & & \\
0 \to \quad K^A & \longrightarrow & P^A & \longrightarrow & N^A & \longrightarrow & 0
\end{array}
$$

with exact rows. Hence, by the Snake Lemma, $\eta_{R^A,K}$ is an epimorphism, and so K is finitely generated. ∎

Now we are in position to determine just when $\mathrm{Gen}(V_R)$ is closed under direct products.

Proposition 1.2.3. *The following statements about a module V_R with $S = \mathrm{End}(V_R)$ are equivalent:*

(a) $\mathrm{Gen}(V_R)$ contains V^A for all sets A;
(b) $\mathrm{Gen}(V_R)$ is closed under direct products;
(c) ${}_S V$ is finitely generated.

Proof. A module M_R is in $\mathrm{Gen}(V_R)$ if and only if the canonical trace mapping $\nu_M : \mathrm{Hom}_R(V, M) \otimes_S V \to M$ is epic. For any set A we have the commutative diagram

$$
\begin{array}{ccc}
\mathrm{Hom}_R(V, V^A) \otimes_S V & \overset{\cong}{\to} & \mathrm{Hom}_R(V, V)^A \otimes_S V \\
\downarrow \nu_{V^A} & & \downarrow = \\
V^A & \overset{\eta_{S^A,V}}{\longleftarrow} & S^A \otimes_S V,
\end{array}
$$

so $(a) \Leftrightarrow (c)$ follows from Lemma 1.2.2.

$(b) \Rightarrow (a)$ is clear. For $(c) \Rightarrow (b)$, assume that ${}_S V$ is finitely generated and $\{M_\alpha\}_{\alpha \in A}$ belong to $\mathrm{Gen}(V_R)$. Then the composite of the canonical homomorphisms

$$
\mathrm{Hom}_R(V, \Pi_A M_\alpha) \otimes_S V \cong (\Pi_A \mathrm{Hom}_R(V, M_\alpha)) \otimes_S V
$$
$$
\overset{\eta}{\longrightarrow} \Pi_A(\mathrm{Hom}_R(V, M_\alpha) \otimes_S V)
$$
$$
\overset{\Pi_A \nu_{M_\alpha}}{\longrightarrow} \Pi_A M_\alpha
$$

is epic by Lemma 1.2.1 and this composite is $\nu_{\Pi M_\alpha}$. ∎

Next we obtain a mapping, in addition to the trace map, that determines whether a module belongs to $\mathrm{Gen}(V^A)$.

Lemma 1.2.4. *Let $i : R_R \to V_R$ be a homomorphism. If*

$$\operatorname{Hom}_R(i, M) : \operatorname{Hom}_R(V, M) \to \operatorname{Hom}_R(R, M)$$

is an epimorphism, then $M \in \operatorname{Gen}(V_R)$.

Proof. Let $\varphi : \operatorname{Hom}_R(R, M) \to M$ be the canonical isomorphism. Then, by hypothesis, for each $m \in M$ there is an $f_m \in \operatorname{Hom}_R(V, M)$ such that $m = \varphi(\operatorname{Hom}_R(i, M)(f_m)) = \varphi(f_m \circ i) = f_m(i(1))$, so $m \in \operatorname{Tr}_V(M)$. ∎

Let V_R be a fixed module with $S = \operatorname{End}(V_R)$. Let $\{x_\alpha\}_{\alpha \in A}$ be a generating set for ${}_S V$ and define $i : R_R \to V_R^A$ via $i(r) = (x_\alpha r)_{\alpha \in A}$. Plainly, $\operatorname{Ker}(i) = \operatorname{Ann}_R(V)$. Thus we have the exact sequence

$$0 \to \operatorname{Ann}_R(V) \to R_R \xrightarrow{i} V_R^A \to (V^A/i(R))_R \to 0.$$

For any $M \in \operatorname{Mod}–R$, denote by i_M^* the composite of the homomorphisms

$$\operatorname{Hom}_R(V^A, M) \xrightarrow{\operatorname{Hom}(i,M)} \operatorname{Hom}_R(R, M) \xrightarrow{\cong} M.$$

Proposition 1.2.5. *$M \in \operatorname{Gen}(V_R^A)$ if and only if i_M^* is epic. In particular, if V_R is finitely generated over its endomorphism ring (and A is taken to be finite), then i_M^* is epic if and only if $M \in \operatorname{Gen}(V_R)$.*

Proof. Denote the class of $M \in \operatorname{Mod}-R$ for which i_M^* is epic by $\mathcal{E}$. We first claim that $V_R \in \mathcal{E}$ and $\mathcal{E}$ is closed under epimorphic images. If $v \in V$, let $v = \Sigma_{j=1}^k s_{\alpha_j} x_{\alpha_j}$. Define $f \in \operatorname{Hom}_R(V^A, V)$ via $f((v_\alpha)) = \Sigma_{j=1}^k s_{\alpha_j} v_{\alpha_j}$. Then $i_V^*(f) = (f \circ i)(1) = v$. For the second assertion suppose that $M \xrightarrow{\eta} L$ is epic in R-Mod where i_M^* is also epic. Then, since $\eta \circ i_M^* = i_L^* \circ \operatorname{Hom}(V^A, \eta)$, i_L^* is also epic and our claim is proved.

Next we note that $\mathcal{E}$ is closed under arbitrary direct sums and products. Plainly, $\mathcal{E}$ is closed under arbitrary direct products and hence under finite direct sums. Let $M = \oplus_{\beta \in B} M_\beta$, where each $M_\beta \in \mathcal{E}$, and let $m \in M$. There is a finite subset B_0 of B such that if κ is the canonical inclusion $M_0 = \oplus_{\beta \in B_0} M_\beta \hookrightarrow M$, $m = \kappa(m')$, with $m' \in M_0$. But then there exists $f \in \operatorname{Hom}_R(V^A, M_0)$ such that $m = \kappa(m') = \kappa(i_{M_0}^*(f)) = i_M^*(\operatorname{Hom}(V^A, \kappa)(f))$.

Now, $\operatorname{Gen}(V^A) \subseteq \mathcal{E}$ follows from what we have proved thus far. The reverse inclusion follows immediately from Lemma 1.2.4. ∎

Since the class of modules M for which i_M^* is epic is clearly closed under direct products, Proposition 1.2.5 implies that $\operatorname{Gen}(V^A)$ is closed under direct products; hence, we have the following corollary by Proposition 1.2.3.

Corollary 1.2.6. *For any* V_R*, if* V *is generated over its endomorphism ring by* $|A|$ *elements, then* V^A *is finitely generated over its endomorphism ring.*

A module V_R is *small* if, as is the case for a finitely generated module, $\mathrm{Hom}_R(V, \oplus_A M_\alpha) \cong \oplus_A \mathrm{Hom}_R(V, M_\alpha)$, canonically, for all $\{M_\alpha\}_A$ in Mod-R. A module V_R is *self-small* if $\mathrm{Hom}_R(V, V^{(A)}) \cong \mathrm{Hom}_R(V, V)^{(A)}$, canonically, for all sets A. This notion is a key element in the proof of the following proposition due to J. Trlifaj [78].

Proposition 1.2.7. *If* $\mathrm{Hom}_R(V, \_)$ *commutes with direct limits (with directed index sets) of modules in* $\mathrm{Gen}(V_R)$*, then* V_R *is finitely generated.*

Proof. First we note that, since $V^{(A)} = \varinjlim V^{(F)}$ such that F is a finite subset of A [69, pp. 44–45], we have, by hypothesis, $\mathrm{Hom}_R(V, V^{(A)}) = \mathrm{Hom}_R(V, \varinjlim V^{(F)}) \cong \varinjlim \mathrm{Hom}_R(V, V^{(F)}) \cong \mathrm{Hom}_R(V, V)^{(A)}$; thus, V is self-small. Let $V = \sum_A x_\alpha R$ and let $\iota_\alpha : V \to V^{(A)}$ ($\alpha \in A$) be the canonical injections. Then, identifying $x_\alpha = \iota_\alpha x_\alpha$, since $V \cong (\oplus_A \iota_\alpha x_\alpha R)/K$, we have a monomorphism

$$\varphi : V \to V^{(A)}/K$$

with $K \leq \oplus_A x_\alpha R$ and

$$\varphi : x_\alpha \mapsto x_\alpha + K.$$

Then, letting $\{K_i\}_{i \in I}$ denotes the finitely generated submodules of K, with canonical epimorphisms $\gamma_i : V^{(A)}/K_i \to V^{(A)}/K$ ($i \in I$),

$$(V^{(A)}/K, \{\gamma_i\}_I) = \varinjlim V^{(A)}/K_i.$$

Now, by hypothesis,

$$(\mathrm{Hom}_R(V, V^{(A)}/K), \{\mathrm{Hom}_R(V, \gamma_i)\}_I) = \varinjlim \mathrm{Hom}_R(V, V^{(A)}/K_i)$$

so that (see [69, Theorem 2.17])

$$\mathrm{Hom}_R(V, V^{(A)}/K) = \cup_I \mathrm{Im} \, \mathrm{Hom}_R(V, \gamma_i).$$

Thus there is an $i \in I$ and a $\varphi_i \in \mathrm{Hom}_R(V, V^{(A)}/K_i)$ with

$$\varphi = \gamma_i \varphi_i.$$

There is a finite set $F \subseteq A$ such that $K_i \subseteq V^{(F)}$, and hence

$$V^{(A)}/K_i = V^{(F)}/K_i \oplus V^{(A \backslash F)}.$$

So since V is self-small

$$\operatorname{Im} \varphi_i \subseteq V^{(H)}/K_i$$

for some finite set H with $F \subseteq H \subseteq A$. Now we have

$$\varphi(V) \subseteq \gamma_i(V^{(H)}/K_i) = (V^{(H)} + K)/K,$$

so, for each $\alpha \in A$, there is a $v_\alpha \in V^{(H)}$ such that

$$v_\alpha + K = \varphi(x_\alpha) = x_\alpha + K.$$

But then

$$v_\alpha \in V^{(H)} \cap (\oplus_A x_\alpha R) = \oplus_H x_\alpha R,$$

and we have

$$\operatorname{Im} \varphi \subseteq (\oplus_H x_\alpha R + K)/K \subseteq \operatorname{Im} \varphi;$$

thus, $V \cong \operatorname{Im} \varphi$ is finitely generated. ∎

Another closure property of $\operatorname{Gen}(V_R)$ forces V to be flat over its endomorphism ring.

Proposition 1.2.8. *Suppose $V \in \text{Mod-}R$ and $S = \operatorname{End}(V_R)$. If $\operatorname{Gen}(V_R)$ is closed under submodules, then $_S V$ is flat.*

Proof. Recall [1, Lemma 19.19] that $_S V$ is flat if and only if for every relation

$$\sum_{i=1}^m s_i x_i = 0 \qquad (s_i \in S, \ x_i \in V)$$

there exist $y_j \in V, \sigma_{ij} \in S, \quad 1 \le i \le m, 1 \le j \le n,$ such that for all $1 \le i \le m$ and $1 \le j \le n$

$$\sum_{j=1}^n \sigma_{ij} y_j = x_i \text{ and } \sum_{i=1}^m s_i \sigma_{ij} = 0.$$

So suppose we do have

$$\sum_{i=1}^m s_i x_i = 0 \qquad (s_i \in S, \ x_i \in V);$$

let $\pi_j : V^m \to V, \quad 1 \le j \le m,$ be the canonical projections and let

$$K = \operatorname{Ker} d$$

where d is the homomorphism $d : V^{(m)} \to V$ defined by

$$d : z \mapsto \sum_{i=1}^{m} s_i \pi_i(z), \quad z \in V^{(m)}.$$

Then $x = (x_1, \ldots, x_m) \in K$ and so, since V generates K, there exist

$$f_j : V \to K, \quad \text{and} \quad y_j \in V, \quad 1 \leq j \leq n$$

such that

$$x = \sum_{j=1}^{n} f_j y_j.$$

Now let

$$\sigma_{ij} = \pi_i f_j \in S, \quad 1 \leq i \leq m, \ 1 \leq j \leq n,$$

to obtain

$$\sum_{j=1}^{n} \sigma_{ij} y_j = \pi_j(x) = x_i, \quad 1 \leq i \leq m,$$

and for each $u \in V$

$$\sum_{i=1}^{m} s_i \sigma_{ij} u = \sum_{i=1}^{m} s_i \pi_i f_j u = d(f_j u) = 0, \quad 1 \leq j \leq n. \quad \blacksquare$$

1.3. Add(V_R) and Prod(V_R)

We denote the subcategories of Mod-R consisting of all direct summands of a direct sum, respectively, a direct product, of copies of a module V_R by *Add(V_R)*, respectively, by *Prod(V_R)*.

According to [1, Theorems 19.20 and 28.4], if S is a left coherent right perfect ring, then every direct product of projective right S-modules is projective, that is, belongs to Add(S_S). (This result and its converse are due to S. Chase [12], who also proved that if every direct product of copies of S_S is projective, then S is a left coherent right perfect ring.) On the other hand we have

Lemma 1.3.1. *If S is a left coherent right perfect ring, then every projective right S-module belongs to* Prod(S_S).

Proof. Letting $J = J(S)$, suppose that P_S is projective and $P/PJ = \oplus_{\alpha \in A} T_\alpha$ with each T_α simple. Let $Q = S_S^A$. Then $QJ \leq J^A$ and $\oplus_{\alpha \in A} T_\alpha$ is isomorphic to a direct summand of $(S/J)^A \cong S^A/J^A$, and so there is an epimorphism $Q \to P/PJ$. Thus by [1, Lemma 17.17] P, the projective cover of P/PJ, is isomorphic to a direct summand of Q. ∎

This last lemma and the paragraph preceding it tell us that, if S is a left coherent right perfect ring, then $\mathrm{Add}(S_S) = \mathrm{Prod}(S_S)$.

Proposition 1.3.2. *Let V_R be a self-small module with $\mathrm{End}(V_R) = S$. If S is left coherent and right perfect, and ${}_SV$ is finitely presented, then $\mathrm{Prod}(V_R) = \mathrm{Add}(V_R)$.*

Proof. Since V is self-small

$$\mathrm{Hom}_R(V, \_) : \mathrm{Add}(V_R) \rightleftarrows \mathrm{Add}(S_S) : (\_ \otimes_S V)$$

is an equivalence of categories. But $\mathrm{Hom}_R(V, \_)$ commutes with direct products and, by Lemma 1.2.1, so does $(\_ \otimes_S V)$. Thus the proposition follows from the fact that $\mathrm{Add}(S_S) = \mathrm{Prod}(S_S)$. ∎

A ring R is an *artin algebra* if its center K is an artinian ring and R is finitely generated as a K-module. Any finitely generated module over an artin algebra is finitely generated over its endomorphism ring, which is also an artin algebra. Thus we have

Corollary 1.3.3. *If V_R is a finitely generated module over an artin algebra R, then $\mathrm{Prod}(V_R) = \mathrm{Add}(V_R)$.*

Note that we have only used one implication of Chase's theorem. Using his full theorem, H. Krause and M. Saorín showed in [53] that a self-small module V_R with $S = \mathrm{End}(V_R)$ has $\mathrm{Add}(V_R)$ closed under direct products if and only if S is a left coherent right perfect ring and ${}_SV$ is finitely presented. In view of Proposition 1.3.2 this is equivalent to $\mathrm{Prod}(V_R) = \mathrm{Add}(V_R)$.

1.4. Torsion Theory

Definition 1.4.1. If $\mathcal{C}$ is an abelian category, a *torsion theory* in $\mathcal{C}$ is a pair of classes of objects $(\mathcal{T}, \mathcal{F})$ of $\mathcal{C}$ such that

(1) $\mathcal{T} = \{T \in \mathcal{C} \mid \mathrm{Hom}_{\mathcal{C}}(T, F) = 0 \text{ for all } F \in \mathcal{F}\}$,
(2) $\mathcal{F} = \{F \in \mathcal{C} \mid \mathrm{Hom}_{\mathcal{C}}(T, F) = 0 \text{ for all } T \in \mathcal{T}\}$,

(3) for each $X \in C$ there is a subobject T of X such that

$$T \in \mathcal{T} \quad \text{and} \quad X/T \in \mathcal{F}.$$

When this is the case, the objects in $\mathcal{T}$ are called *torsion* objects, the elements of $\mathcal{F}$ are called *torsion-free* objects, and if the object T of (3) is unique, we denote it by $\tau(X)$ and call it the *torsion subobject* of X.

Suppose that C is a full subcategory of Mod-R that is closed under submodules, epimorphic images, extensions, direct sums, and direct products. If $(\mathcal{T}, \mathcal{F})$ is a torsion theory in C, then it follows that $\mathcal{T}$ is closed under epimorphic images and direct sums, $\mathcal{F}$ is closed under submodules and direct products, and both are closed under extensions. A class $\mathcal{T}$ $(\mathcal{F})$ of modules in C with these closure properties is called a *torsion (torsion-free) class* in C. Then one easily verifies

Proposition 1.4.2. *Let C be a full subcategory of* Mod-R *that is closed under submodules, epimorphic images, extensions, direct sums, and direct products.*

(1) *If $\mathcal{T}$ is a torsion class in C, then $(\mathcal{T}, \mathcal{F})$ is a torsion theory in C, where $\mathcal{F} = \{F \in C \mid \operatorname{Hom}_C(T, F) = 0 \text{ for all } T \in \mathcal{T}\}$.*
(2) *If $\mathcal{F}$ is a torsion-free class in C, then $(\mathcal{T}, \mathcal{F})$ is a torsion theory in C, where $\mathcal{T} = \{T \in C \mid \operatorname{Hom}_C(T, F) = 0 \text{ for all } F \in \mathcal{F}.$*

Dual to Gen($\mathcal{V}$), if $\mathcal{V}$ is a class of R-modules, Cogen (V) (cogen($\mathcal{V}$)) consists of the R-modules that embed in (finite) direct products of modules isomorphic to members of $\mathcal{V}$, and the *reject of $\mathcal{V}$ in M* is $\operatorname{Rej}_\mathcal{V}(M)$, the intersection of the kernels of all maps from M into members of $\mathcal{V}$.

Proposition 1.4.3. *Let $(\mathcal{T}, \mathcal{F})$ be a torsion theory in* Mod-R *and M a module in* Mod-R. *Then*

$$\operatorname{Tr}_\mathcal{T}(M) = \operatorname{Rej}_\mathcal{F}(M).$$

Proof. That $\operatorname{Tr}_\mathcal{T}(M) \subseteq \operatorname{Rej}_\mathcal{F}(M)$ follows from $\operatorname{Hom}_R(T, F) = 0$ whenever $T \in \mathcal{T}$ and $F \in \mathcal{F}$. But since $\operatorname{Tr}_\mathcal{T}(M) \in \mathcal{T}$ and $\mathcal{T}$ is closed under extensions, $M/\operatorname{Tr}_\mathcal{T}(M) \in \mathcal{F}$ and hence $\operatorname{Rej}_\mathcal{F}(M) \subseteq \operatorname{Tr}_\mathcal{T}(M)$. ∎

If $(\mathcal{T}, \mathcal{F})$ is a torsion theory in Mod-R, we let

$$\tau_\mathcal{T}(M) = \operatorname{Tr}_\mathcal{T}(M) = \operatorname{Rej}_\mathcal{F}(M)$$

and call it the *torsion submodule* of M. Then every module in Mod-R admits an exact sequence

$$0 \to \tau_T(M) \longrightarrow M \longrightarrow M/\tau_T(M) \to 0$$

with $\tau_T(M)$ simultaneously the largest submodule of M belonging to $\mathcal{T}$ and the smallest submodule of M such that $M/\tau_T(M)$ belongs to $\mathcal{F}$.

We shall meet torsion theories like those in the following proposition in later sections.

Proposition 1.4.4. *If* $\mathrm{Gen}(V_R) \subseteq V^{\perp}$, *then* $(\mathrm{Gen}(V_R), \mathrm{Ker}\,\mathrm{Hom}_R(V, \_))$ *is a torsion theory in* Mod-R.

Proof. If $0 \to M_1 \longrightarrow X \longrightarrow M_2 \to 0$ is exact with $M_1,\ M_2 \in \mathrm{Gen}(V_R) \subseteq V^{\perp}$, letting $S = \mathrm{End}(V_R)$, we obtain a commutative diagram with exact rows

$$\mathrm{Hom}_R(V, M_1) \otimes_S V \to \mathrm{Hom}_R(V, X) \otimes_S V \to \mathrm{Hom}_R(V, M_2) \otimes_S V \to 0$$
$$\begin{array}{ccccc} & \nu_{M_1} \downarrow & & \nu_X \downarrow & & \nu_{M_2} \downarrow \\ 0 \to & M_1 & \to & X & \to & M_2 \end{array}$$

in which the trace maps ν_{M_1} and ν_{M_2} are epimorphisms. But then, by the Snake Lemma, so is ν_X, and Proposition 1.4.2 applies. ∎

2

Representable Equivalences

We are concerned with equivalences and dualities between subcategories of the categories of modules over rings. Henceforth, by "subcategory" we shall mean full subcategory that is closed under isomorphic images, and all functors between categories of modules are assumed to be additive.

Suppose C and D are subcategories of Mod-R and Mod-S, respectively. A functor $H : C \to D$ is an *equivalence* if there is a functor $T : D \to C$ such that $T \circ H$ and $H \circ T$ are naturally isomorphic to the identity functors 1_C and 1_D, respectively. When this is the case we write $C \approx D$. By Theorem A.3.4 these natural isomorphisms can be taken to be of the form $\mu : TH \to 1_C$ and $\theta : 1_D \to HT$ where $H\mu \circ \theta H = 1_H$ and $\mu T \circ T\theta = 1_T$. That is, μ and θ, an arrow of adjunction and its quasi-inverse, establish T as a left adjoint of H (see Appendix A). If $_S V_R$ is a bimodule, then we have functors

$$\mathrm{Hom}_R(V, \_) : \mathrm{Mod}\text{-}R \rightleftarrows \mathrm{Mod}\text{-}S : (\_ \otimes_S V),$$

and we say that the equivalence $H : C \rightleftarrows D : T$ is *representable by* $_S V_R$ if H and T are naturally isomorphic to the restrictions of these functors, that is,

$$H \cong \mathrm{Hom}_R(V, \_)|C \quad \text{and} \quad T \cong (\_ \otimes_S V)|D.$$

In this case we shall make the identifications

$$H = \mathrm{Hom}_R(V, \_) \quad \text{and} \quad T = (\_ \otimes_S V),$$

and then by Theorem A.2.2 the canonical natural transformations ν and η defined below are natural isomorphisms when restricted to C and D, respectively.

2.1. Adjointness of $\mathrm{Hom}_R(V, \_)$ and $\_ \otimes_S V$

If $_S V_R$ is a bimodule, then for any $M \in \mathrm{Mod}\text{-}R$ and $N \in \mathrm{Mod}\text{-}S$ there is an isomorphism

$$\alpha = \alpha_{N,M} : \mathrm{Hom}_S(N, \mathrm{Hom}_R(V, M)) \to \mathrm{Hom}_R((N \otimes_S V), M)$$

with, for $\delta \in \mathrm{Hom}_S(N, \mathrm{Hom}_R(V, M))$, $n \in N$, $v \in V$, and $\gamma \in \mathrm{Hom}_R$ $((N \otimes_S V), M)$,

$$\alpha(\delta) : n \otimes v \mapsto \delta(n)(v) \quad \text{and} \quad a^{-1}(\gamma)(n) : v \mapsto \gamma(n \otimes v).$$

The isomorphism α is natural in both M and N, that is, α defines two natural transformations

$$\alpha : \mathrm{Hom}_S(N, \mathrm{Hom}_R(V, \_)) \rightarrow \mathrm{Hom}_R((N \otimes_S V), \_)$$

and

$$\alpha : \mathrm{Hom}_S(\_, \mathrm{Hom}_R(V, M)) \rightarrow \mathrm{Hom}_R((\_ \otimes_S V), M).$$

In this chapter, when $_S V_R$ is a given bimodule, we shall write

$$HM = \mathrm{Hom}_R(V, M) \quad \text{and} \quad TN = N \otimes_S V$$

for $M \in \mathrm{Mod}\text{-}R$ and $N \in \mathrm{Mod}\text{-}S$. Thus the functor $T : \mathrm{Mod}\text{-}S \rightarrow \mathrm{Mod}\text{-}R$ is a left adjoint of the functor $H : \mathrm{Mod}\text{-}R \rightarrow \mathrm{Mod}\text{-}S$. Moreover, associated with this adjunction there are the canonical natural transformations

$$\nu : TH \rightarrow 1_{\mathrm{Mod}\text{-}R} \quad \text{and} \quad \eta : 1_{\mathrm{Mod}\text{-}S} \rightarrow HT$$

with, for each $\varphi \in HM$, $v \in V$ and $n \in N$,

$$\nu_M : \varphi \otimes v \mapsto \varphi(v) \quad \text{and} \quad \eta_N(n) : v \mapsto n \otimes v.$$

Direct calculations verify that

$$H(\nu_M) \circ \eta_{HM} = 1_{HM} \quad \text{and} \quad \nu_{TN} \circ T(\eta_N) = 1_{TN}. \tag{2.1}$$

We shall say that M is ν-*reflexive* if ν_M is an isomorphism and that N is η-*reflexive* if η_N is an isomorphism. The preceding display easily yields

Proposition 2.1.1. *For all* $M \in \mathrm{Mod}\text{-}R$, *and all* $N \in \mathrm{Mod}\text{-}S$,

    (1) η_{HM} *is a split monomorphism and* ν_{TN} *is a split epimorphism;*
    (2) If M *is* ν-*reflexive, then* HM *is* η-*reflexive; if* N *is* η-*reflexive, then* TN *is* ν-*reflexive.*

Thus H and T yield an equivalence between the category of ν-reflexive modules in $\mathrm{Mod}\text{-}R$ and the category of η-reflexive modules in $\mathrm{Mod}\text{-}S$.
    Note that

$$\mathrm{Im}\, \nu_M = \mathrm{Tr}_V(M).$$

On the other hand, given an injective cogenerator C_R, let

$$V_S^* = \text{Hom}_R(_S V_R, C_R).$$

Then, according to [1, Exercise 19.20],

$$\text{Ker } \eta_N = \text{Ann}_N(_S V) = \text{Rej}_{V^*}(N).$$

Thus we have

Lemma 2.1.2. *If $M \in$ Mod-R and $N \in$ Mod-S, then*

(1) ν_M is an epimorphism if and only if $M \in \text{Gen}(V_R)$;

(2) η_N is a monomorphism if and only if $N \in \text{Cogen}(V_S^*)$.

We let $\text{Pres}(V_R)$ denote the class of R-modules that are *presented* by V in the sense that there is an exact sequence

$$V^{(B)} \longrightarrow V^{(A)} \longrightarrow M \to 0$$

for some sets A and B. Dually, $M \in \text{Copres}(W_R)$ if it is *copresented* by W in the sense that there is an exact sequence

$$0 \to M_R \longrightarrow W^A \longrightarrow W^B.$$

Applying T and H to exact sequences $S^{(B)} \longrightarrow S^{(A)} \xrightarrow{g} N_S \to 0$ and $0 \to M_R \longrightarrow C^A \longrightarrow C^B$, with C_R an injective cogenerator, we obtain

Lemma 2.1.3. *If $M \in$ Mod-R and $N \in$ Mod-S, then*

(1) $TN \in \text{Pres}(V_R) \subseteq \text{Gen}(V_R)$;

(2) $HM \in \text{Copres}(V_S^*) \subseteq \text{Cogen}(V_S^*)$.

Also we note that

Proposition 2.1.4. *The module S_S is η-reflexive if and only if $S \cong \text{End}(V_R)$ canonically, and if so, then V_R is ν-reflexive.*

2.2. Weak ∗-modules

Suppose that $\mathcal{C}$ and $\mathcal{D}$ are subcategories of Mod-R and Mod-S, respectively, with an equivalence $H : \mathcal{C} \rightleftarrows \mathcal{D} : T$. If $S_S \in \mathcal{D}$, we have natural isomorphisms

$$H(\_) \cong \text{Hom}_S(S, H(\_)) \cong \text{Hom}_R(T(S), \_).$$

Hence, letting $_S V_R = T(_S S_S)$ be an S–R bimodule in the canonical way, H is isomorphic to $\text{Hom}_R(V, \_)$. Suppose, in addition, that $\_ \otimes_S V : \mathcal{D} \to \mathcal{C}$. Then by [1, Exercise 20.7], T is naturally isomorphic to $\_ \otimes_S V$, and thus the equivalence $H : \mathcal{C} \rightleftarrows \mathcal{D} : T$ is representable by $_S V_R$. Also note that in this setting we have $HT(S) = \text{End}(V_R)$.

Throughout the remainder of this chapter, we shall assume that we are dealing with a given bimodule $_S V_R$ and that H, T, ν, η and V^* are as in Section 2.1.

As we shall see, most of the interesting equivalences between module categories are ones between $\text{Gen}(V_R)$ and $\text{Cogen}(V_S^*)$, namely, those that are induced by a so-called ∗-module. En route, here we shall investigate equivalences between $\text{Pres}(V_R)$ and $\text{Cogen}(V_S^*)$.

Lemma 2.2.1. *(1) Suppose that* $0 \to K \longrightarrow X \xrightarrow{f} M \to 0$ *is exact in Mod-R, that X is ν-reflexive, and Hf is epic. Then M is ν-reflexive if and only if* $K \in \text{Gen}(V_R)$.

(2) Suppose that $0 \to N \xrightarrow{g} Y \longrightarrow L \to 0$ *is exact in Mod-S, that Y is η-reflexive, and that Tg is monic. Then N is η-reflexive if and only if* $L \in \text{Cogen}(V_S^*)$.

Proof. (1) Apply the Snake Lemma to the commutative diagram

$$THK \to THX \to THM \to 0$$
$$\downarrow \qquad \downarrow \qquad \downarrow$$
$$0 \to \quad K \ \to \ X \ \to \ M \ \to 0.$$

(2) Apply the Snake lemma to

$$0 \to \quad N \ \to \ Y \ \to \ L \ \to 0$$
$$\downarrow \qquad \downarrow \qquad \downarrow$$
$$0 \to HTN \to HTY \to HTL \qquad \blacksquare$$

The next lemma provides a connection between reflexivity and exactness of H and T.

Lemma 2.2.2. *(1) Suppose* $0 \to K \longrightarrow X \xrightarrow{f} M \to 0$ *is exact in Mod-R. If X is ν-reflexive and* $K \in \text{Gen}(V_R)$, *then Hf is epic if and only if* $\text{Im } Hf$ *is η-reflexive.*

(2) Suppose $0 \to N \xrightarrow{g} Y \longrightarrow L \to 0$ *is exact in Mod-S. If Y is η-reflexive and* $L \in \text{Cogen}(V_S^*)$, *then Tg is monic if and only if* $\text{Im } Tg$ *is ν-reflexive.*

Proof. (1) From the exact sequence

$$0 \to HK \longrightarrow HM \xrightarrow{p} \operatorname{Im} Hf \to 0$$

and the embedding

$$0 \to \operatorname{Im} Hf \xrightarrow{i} HM,$$

we obtain a commutative diagram

$$
\begin{array}{ccccccc}
& & THK & \longrightarrow & THX & \xrightarrow{Tp} & T(\operatorname{Im} Hf) \to 0 \\
& & v_K \downarrow & & v_X \downarrow & & \alpha \downarrow \\
0 \to & & K & \longrightarrow & X & \xrightarrow{f} & M & \to 0
\end{array}
$$

where $\alpha = v_M \circ T(i)$. Here, by the Five Lemma, α, and hence $H(\alpha)$, is an isomorphism. Also

$$
\begin{aligned}
i &= 1_{HM} \circ i \\
&= H(v_M) \circ \eta_{HM} \circ i \\
&= H(v_M) \circ HT(i) \circ \eta_{\operatorname{Im} Hf} \\
&= H(\alpha) \circ \eta_{\operatorname{Im} Hf}
\end{aligned}
$$

where the third equality is due to the naturalness of η :

$$
\begin{array}{ccc}
\operatorname{Im} Hf & \xrightarrow{i} & HM \\
\eta_{\operatorname{Im} Hf} \downarrow & & \eta_{HM} \downarrow \\
HT(\operatorname{Im} Hf) & \xrightarrow{HT(i)} & HTHM.
\end{array}
$$

Thus Hf is epic if and only if i is an isomorphism if and only if $\eta_{\operatorname{Im} Hf}$ is an isomorphism.

(2) From the exact sequence

$$0 \to \operatorname{Im} Tg \xrightarrow{j} TY \longrightarrow TL \to 0$$

and the epimorphism

$$TN \xrightarrow{q} \operatorname{Im} Tg \longrightarrow 0$$

we obtain a commutative diagram

$$
\begin{array}{ccccccc}
0 \to & N & \xrightarrow{g} & Y & \to & L & \to 0 \\
& \beta \downarrow & & \eta_Y \downarrow & & \eta_L \downarrow \\
0 \to & H(ImTg) & \xrightarrow{Hj} & HT(Y) & \to & HT(L)
\end{array}
$$

where $\beta = H(q) \circ \eta_N$. Here, by the Five Lemma, β, and hence $T(\beta)$, is an isomorphism. Also

$$
\begin{aligned}
q &= q \circ 1_{TN} \\
&= q \circ \nu_{TN} \circ T(\eta_N) \\
&= \nu_{\operatorname{Im} Tg} \circ TH(q) \circ T(\eta_N) \\
&= \nu_{\operatorname{Im} Tg} \circ T(\beta)
\end{aligned}
$$

where the third equality is due to the naturalness of ν. Thus $\operatorname{Im} Tg$ is ν-reflexive if and only if q is an isomorphism if and only if Tg is monic. ∎

Recall that M is self-small if $\operatorname{Hom}_R(M, M^{(A)}) \cong \operatorname{Hom}_R(M, M)^{(A)}$, canonically, for all sets A.

Lemma 2.2.3. *If $S = \operatorname{End}(V_R)$, then V_R is self-small if and only if $S_S^{(A)}$ is η-reflexive for all sets A. Moreover, if this is the case, then $V^{(A)}$ is ν-reflexive.*

Proof. Assuming that V is self-small,

$$
S^{(A)} \cong \operatorname{Hom}_R(V, V^{(A)}) \cong HT(S^{(A)}),
$$

and these isomorphisms compose to $\eta_{S^{(A)}}$. Conversely, if $S_S^{(A)}$ is η-reflexive, then

$$
\operatorname{Hom}_R(V, V^{(A)}) = HT(S^{(A)}) \cong S^{(A)} = \operatorname{Hom}_R(V, V)^{(A)}.
$$

The last statement follows, since $T(S^{(A)}) \cong V_R^{(A)}$. ∎

Now we are prepared to prove the main result of this section.

Proposition 2.2.4. *If $S = \operatorname{End}(V_R)$, the following are equivalent:*

(a) $H : \operatorname{Pres}(V_R) \rightleftarrows \operatorname{Cogen}(V_S^*) : T$ *is an equivalence;*
(b) N_S *is η-reflexive whenever η_N is a monomorphism;*
(c) η_N *is epic for all $N \in \operatorname{Mod-}S$;*
(d) V_R *is self-small, and for each exact sequence $0 \to K \longrightarrow V^{(A)} \overset{f}{\longrightarrow} M \to 0$ in $\operatorname{Mod-}R$, Hf is epic if $K \in \operatorname{Gen}(V_R)$.*

Proof. (a) $\Rightarrow$ (b). This is obvious, in view of Lemma 2.1.2.

(b) $\Rightarrow$ (d). Since $S_S^{(A)} = (HV)^{(A)} \in \operatorname{Cogen}(V_S^*)$, assuming (b), Lemmas 2.1.2 and 2.2.3 imply V_R is self-small and $V^{(A)}$ is ν-reflexive. Also, by Lemma 2.1.3, $\operatorname{Im} Hf \subseteq HM \in \operatorname{Cogen}(V_S^*)$, and so by Lemma 2.1.2 $\eta_{\operatorname{Im} Hf}$ is monic. Thus by (b) and Lemma 2.2.2 Hf is epic if $K \in \operatorname{Gen}(V_R)$.

$(d) \Rightarrow (a)$. As noted above, $HM \in \mathrm{Cogen}(V_S^*)$. Since V_R is self-small, if $M \in \mathrm{Pres}(V_R)$, then, assuming K is in $\mathrm{Gen}(V)$ in (d), M is ν-reflexive by Lemma 2.2.1. Suppose η_N is monic. Since $N \in \mathrm{Pres}(S_S)$, we have an exact sequence $0 \to K \longrightarrow V^{(A)} \overset{Tg}{\longrightarrow} TN \to 0$ with $K \in \mathrm{Gen}(V_R)$. Thus, by (d), HTg is epic, and we have a commutative diagram with exact rows

$$
\begin{array}{ccc}
S^{(A)} & \overset{g}{\longrightarrow} & N \quad \to 0 \\
\eta_{S^{(A)}} \downarrow & & \eta_N \downarrow \\
HT(S^{(A)}) & \overset{HTg}{\longrightarrow} & HTN \to 0.
\end{array}
$$

So since $S^{(A)}$ is η-reflexive by Lemma 2.2.3, η_N is an isomorphism.

$(b) \Leftrightarrow (c)$. Since $\mathrm{Ann}_N(_SV) = \mathrm{Rej}_{V_S^*}(N) = \mathrm{Ker}\,\eta_N$, we have an exact sequence

$$
0 \to \mathrm{Ann}_N(_SV) \to N \overset{p}{\longrightarrow} N/\mathrm{Rej}_{V_S^*}(N) \to 0.
$$

It follows that Tp is an isomorphism. Thus, from the commutative diagram

$$
\begin{array}{ccc}
N & \overset{p}{\longrightarrow} & N/\mathrm{Rej}_{V_S^*}(N) \\
\eta_N \downarrow & & \eta_{N/\mathrm{Rej}_{V_S^*}(N)} \downarrow \\
HT(N) & \overset{HTp}{\longrightarrow} & HTN
\end{array}
$$

we see that $\eta_{N/\mathrm{Rej}_{V_S^*}(N)}$ is an isomorphism if and only if η_N is an epimorphism. ∎

An R-module Q is *quasi-projective* if, whenever $Q \to L \to 0$ is exact, so is $\mathrm{Hom}_R(Q, Q) \to \mathrm{Hom}_R(Q, L) \to 0$.

Example 2.2.5. A module V_R induces an equivalence

$$
H : \mathrm{Pres}(V_R) \rightleftarrows \mathrm{Cogen}(V_S^*) : T
$$

if (1) V_R is finitely generated and (quasi-)projective, or (2) V_R is self-small and $\mathrm{Gen}(V_R) \subseteq \mathrm{Ker}\,\mathrm{Ext}_R^1(V, \_)$, since in either case V satisfies (d).

2.3. ∗-modules

Suppose $\mathcal{C}$ and $\mathcal{D}$ are subcategories of Mod-R and Mod-S, respectively, and that $\mathcal{C}$ is closed under direct sums and epimorphisms and $\mathcal{D}$ is closed under direct products and submodules. If $I = \{a \in S \mid Na = 0 \text{ for all } N \in \mathcal{D}\}$, then I is an ideal of S, such that $S/I \in \mathcal{D}$ and $\mathcal{D}$ is a subcategory of Mod-S/I. Thus, supposing further that there is an equivalence $\mathcal{C} \approx \mathcal{D}$, we may assume that $S \in \mathcal{D}$. Then, as discussed at the beginning Section 2.2, we may also assume that the equivalence is of the form

$$H : \mathcal{C} \rightleftarrows \mathcal{D} : T$$

induced by a bimodule $_S V_R = T(S)$ with $S = \text{End}(V_R)$ because an epimorphism $S^{(A)} \to N_S$ yields an epimorphism $V^{(A)} \to N \otimes_S V_R$. Now $\text{Gen}(V_R) \subseteq \mathcal{C} = T(\mathcal{D}) \subseteq \text{Gen}(V_R)$ and $\mathcal{D} = H(\mathcal{C}) \subseteq \text{Cogen}(V_S^*)$ by Lemma 2.1.3. But also $V_S^* = \text{Hom}_R(V, C) = H(Tr_V(C)) \in H(\mathcal{C}) = \mathcal{D}$. Thus $\mathcal{C} = \text{Gen}(V_R)$ and $\mathcal{D} = \text{Cogen}(V_S^*)$, so our equivalence is induced by a *-module in the sense of the following definition.

Definition 2.3.1. A module V_R with $S = \text{End}(V_R)$ is a *-*module* if

$$H : \text{Gen}(V_R) \rightleftarrows \text{Cogen}(V_S^*) : T$$

is an equivalence.

Although the term *-module was coined later, the notion was introduced by C. Menini and A. Orsatti [62] as a generalization of both quasi-progenerators and tilting modules. In addition to these authors, R. Colpi [24] and J. Trlifaj [78] are responsible for most of the results in this section.

From the discussion preceding Definition 2.3.1, we have

Proposition 2.3.2. *Let $\mathcal{C}$ be a subcategory of* Mod-R *that is closed under direct sums and epimorphic images, and let $\mathcal{D}$ be a subcategory of* Mod-S *that is closed under direct products and submodules. Let $I = \mathbf{r}_S(\mathcal{D})$, that is,*

$$I = \{a \in S \mid Na = 0 \text{ for all } N \in \mathcal{D}\}.$$

*Then any equivalence $\mathcal{C} \approx \mathcal{D}$ is representable by a *-module V_R (corresponding to S/I_S under the equivalence) with $S/I \cong \text{End}(V_R)$, $\mathcal{C} = \text{Gen}(V_R)$ and $\mathcal{D} = \text{Cogen}(V_S^*)$*

If V_R is a *-module with $S = \text{End}(V_R)$, then V_R acts as a projective object in $\text{Gen}(V_R)$ and $_S V$ is "flat" in $\text{Cogen}(V_S^*)$.

Proposition 2.3.3. *If V_R is a *-module with $S = \text{End}(V_R)$, then H is exact on short exact sequences in $\text{Gen}(V_R)$, and T is exact on short exact sequences in $\text{Cogen}(V_S^*)$.*

Proof. This follows at once from Lemma 2.2.2. ∎

Lemma 2.3.4. *If $_SW_R$ is a bimodule and Q_R is injective, then for any N_S,*

$$\text{Ext}_S^i(N, \text{Hom}_R(W, Q)) \cong \text{Hom}_R(\text{Tor}_i^S(N, W), Q)$$

for all $i \geq 0$.

Proof. This is [11, Page 120, Proposition 5.1]. ∎

Now we have the necessary tools to give homological characterizations of $\text{Cogen}(V_S^*)$ for a $*$-module V_R.

Proposition 2.3.5. *If V_R is a $*$-module with $S = \text{End}(V_R)$, then*

$$\text{Cogen}(V_S^*) = \text{Ker Tor}_1^S(\_, V) = \text{Ker Ext}_S^1(\_, V^*).$$

Proof. Given a right S-module L, there is an exact sequence $0 \to N \xrightarrow{g} S^{(A)} \xrightarrow{f} L \to 0$ in which $S_S = HV \in \text{Cogen}(V_S^*)$. Thus $S^{(A)}$ and N belong to $\text{Cogen}(V_S^*)$. Now we have an exact sequence

$$0 = \text{Tor}_1^S(S^{(A)}, V) \to \text{Tor}_1^S(L, V) \to TN \xrightarrow{Tg} TS^{(A)} \xrightarrow{Tf} TL \to 0.$$

So if $L \in \text{Cogen}(V_S^*)$, then Tg is monic by Proposition 2.3.3 and hence $L \in \text{Ker Tor}_1^S(\_, V)$. On the other hand, if $\text{Tor}_1^S(L, V) = 0$, then Tg is monic and $L \in \text{Cogen}(V_S^*)$ by (2) of Lemma 2.2.1. The second equality is by Lemma 2.3.4. ∎

This last result and Trilfaj's Proposition 1.2.7 lead to

Theorem 2.3.6. *Every $*$-module V_R is finitely generated.*

Proof. Since $\text{Tor}_n^S(\_, V)$ commutes with direct limits for all $n \geq 0$ [69, Theorem 8.13], if $(\{M_i\}, \{f_{ij}\})$ is a direct system with $M_i \in \text{Gen}(V_R)$, then by Proposition 2.3.5, $\varinjlim H(M_i) \in \text{Cogen}(V_S^*)$. Thus

$$H(\varinjlim M_i) \cong H(\varinjlim TH(M_i)) \cong HT(\varinjlim H(M_i)) \cong \varinjlim H(M_i),$$

so $\varinjlim M_i \cong T(\varinjlim H(M_i)) \in \text{Gen}(V_R)$, and V_R is finitely generated by Proposition 1.2.7. ∎

The following lemma, an early version of which can be found in [8], and its dual version have been applied in several papers on equivalence and duality.

Lemma 2.3.7. *Suppose* $V, M \in$ Mod-R, $S = \text{End}(V_R)$ *and* X *is a set of generators for* $\text{Hom}_R(V, M)_S$. *Then there is an epimorphism*

$$V^{(X)} \overset{\mu}{\to} \text{Tr}_V(M)$$

such that $\text{Hom}_R(V, \mu)$ *is epic. Thus if* $K = \text{Ker}(\mu)$, *there are exact sequences*

$$0 \to \text{Hom}_R(V, K) \to \text{Hom}_R(V, V^{(X)}) \overset{\text{Hom}_R(V,\mu)}{\to} \text{Hom}_R(V, Tr_V(M)) \to 0$$

and

$$0 \to \text{Ext}^1_R(V, K) \to \text{Ext}^1_R(V, V^{(X)}) \overset{\text{Ext}^1_R(V,\mu)}{\to} \text{Ext}^1_R(V, Tr_V(M)).$$

Proof. Since X generates $\text{Hom}_R(V, M)_S$, a straightforward argument shows $(v_f)_X \mapsto \sum_{f \in X} f(v_f)$ defines an epimorphism $V^{(X)} \overset{\mu}{\to} \text{Tr}_V(M)$. To see that $\text{Hom}_R(V, \mu)$ is epic, let $g \in \text{Hom}_R(V, \text{Tr}_V(M)) \subseteq \text{Hom}_R(V, M)$. Then $g = \sum_{i=1}^n f_i s_i$ where $f_i \in X$ and $s_i \in S$. Thus, letting $\iota_f : V \to V^{(X)}$ denote the canonical injections and defining $h \in \text{Hom}_R(V, V^{(X)})$ via $v \mapsto \sum_{i=1}^n \iota_{f_i} s_i v$, we obtain $\text{Hom}_R(V, \mu)(h) = g$. ∎

Now we are able to prove the following characterization of ∗-modules that is largely due to R. Colpi.

Theorem 2.3.8. *The following statements about a module* V_R *with* $S = \text{End}(V_R)$ *are equivalent:*

(a) V_R *is a* ∗*-module;*
(b) v_M *is monic for all* $M \in$ Mod-R *and* η_N *is epic for all* $N \in$ Mod-S;
(c) $\text{Gen}(V_R) = \text{Pres}(V_R)$, V_R *is finitely generated, and* H *is exact on short exact sequences in* $\text{Gen}(V_R)$;
(d) $\text{Gen}(V_R) = \text{Pres}(V_R)$, V_R *is self-small, and for each exact sequence* $0 \to K \longrightarrow V^{(A)} \overset{f}{\longrightarrow} M \to 0$ *in* Mod-R, Hf *is epic if* $K \in \text{Gen}(V_R)$.
(e) V_R *is self-small, and for each exact sequence* $0 \to K \longrightarrow V^{(A)} \overset{f}{\longrightarrow} M \to 0$ *in* Mod-R, Hf *is epic if and only if* $K \in \text{Gen}(V_R)$.

Proof. $(a) \Leftrightarrow (d)$. This follows at once from Proposition 2.2.4 and Lemma 2.1.3.

$(a) \Leftrightarrow (b)$. By Proposition 2.2.4 we need only show that v_M is monic for all $M \in$ Mod-R if v_M is an isomorphism whenever $M \in \text{Gen}(V_R)$. But since $HM = H(\text{Tr}_V(M))$, Hi is an isomorphism where $i : Tr_V(M) \to M$ is the inclusion map. Thus the equivalence follows from the commutative diagram

$$
\begin{array}{ccc}
TH(\mathrm{Tr}_V(M)) & \xrightarrow{THi} & THM \\
\nu_{\mathrm{Tr}_V(M)} \downarrow & & \nu_M \downarrow \\
0 \to \quad \mathrm{Tr}_V(M) & \xrightarrow{\ i\ } & M.
\end{array}
$$

$(a) \Rightarrow (c)$. From (a) and Lemma 2.1.3, $\mathrm{Gen}(V_R) = \mathrm{Pres}(V_R)$. Also V_R is finitely generated by Theorem 2.3.6, and H is exact on $\mathrm{Gen}(V_R)$ by Lemma 2.3.3.

$(c) \Rightarrow (d)$. This is clear.

(a) and $(d) \Rightarrow (e)$. Since $V^{(A)}$ is reflexive by (a), Lemma 2.2.1 shows that Hf epic implies $K \in \mathrm{Gen}(V_R)$.

$(e) \Rightarrow (d)$. Here we need only show that $\mathrm{Gen}(V_R) \subseteq \mathrm{Pres}(V_R)$. But if $M \in \mathrm{Gen}(V_R)$, then by Lemma 2.3.7 there is an epimorphism

$$
V^{(HM)} \xrightarrow{\ f\ } M \to 0
$$

such that Hf is epic, and so by (e) $\mathrm{Ker}\, f \in \mathrm{Gen}(V_R)$ and $M \in \mathrm{Pres}(V_R)$. ∎

2.4. Three Special Kinds of ∗-modules

In this section we consider the three principal types of ∗-modules. They induce equivalences between special categories of modules and historically were the inspiration for the concept.

Definition 2.4.1. A *progenerator* is a finitely generated projective generator.

As we shall prove shortly, progenerators are the modules that induce the so-called *Morita equivalences* between the full categories of modules over two rings, so they are ∗-modules. According to [1, Theorem 17.18], a progenerator V_R with $S = \mathrm{End}(V_R)$ is also a progenerator in S-Mod and is faithfully balanced in the sense that $R \cong \mathrm{End}(_S V)$, canonically.

A module U is *projective relative to* a module M if $\mathrm{Hom}_R(U, f)$ is an epimorphism whenever $M \xrightarrow{f} L \to 0$ is exact. Hence if U is projective relative to itself, U is a quasi-projective module. According to [1, Proposition 16.12], the class of modules that U is projective relative to is closed under submodules, epimorphic images, and finite (arbitrary) direct sums (if U is finitely generated).

Definition 2.4.2. A *quasi-progenerator* is a finitely generated quasi-projective module that generates all its submodules.

Of course the notion of quasi-progenerator is a generalization of that of progenerator. A simple module is always a quasi-progenerator, but most are neither projective nor generators.

Definition 2.4.3. A finitely generated module V_R is a *tilting module* if $\text{Gen}(V_R) = V_R^{\perp}$.

If V_R is a progenerator $\text{Gen}(V_R) = \text{Mod-}R = V_R^{\perp}$, so tilting modules also generalize progenerators. In Theorem 2.4.5 we shall see that progenerators, quasi-progenerators, and tilting modules are indeed ∗-modules, and we shall provide criteria for a ∗-module to be of each of these types. First we need

Lemma 2.4.4. *If V is projective relative to each M_i for $i \in I$ and generates all submodules of each M_i, then V generates every submodule of $\oplus_I M_i$.*

Proof. It suffices to prove that V generates every cyclic submodule of $\oplus_I M_i$, so we may assume that $I = \{1, 2\}$. Suppose that $L \subseteq M_1 \oplus M_2$ with projections π_1 and π_2, and let $x \in L$. Since $\pi_1(x) \in \pi_1(L) \subseteq M_1$, there is an $n \in \mathbb{N}$ and an $f : V^n \to \pi_1(L)$ with $f(y) = \pi_1(x)$ for some $y \in V^n$. By hypothesis, f factors through $\pi_1|_L$, that is, $f = \pi_1 \circ g$ where $g : V^n \to L$. Since $\pi_1 \circ g(y) = f(y) = \pi_1(x)$, $g(y) - x \in L \cap M_2 \subseteq M_2$, so $g(y)$, $x - g(y) \in \text{Tr}_V(L)$; hence, $x \in \text{Tr}_V(L)$ too. ∎

Note that it follows from the next theorem, from [24], [30], and [25], that the class of progenerators is the intersection of the classes of quasi-progenerators and tilting modules.

Theorem 2.4.5. *Let V_R be an R-module. Then*

(1) V_R is a progenerator if and only if V_R is a ∗-module with $\text{Gen}(V_R) = \text{Mod-}R$;

(2) V_R is a quasi-progenerator if and only if V_R is a ∗-module such that $\text{Gen}(V_R)$ is closed under submodules;

(3) V_R is a tilting module if and only if V_R is a ∗-module with $E(R_R) \in \text{Gen}(V_R)$.

Proof. (1) If V_R is a progenerator, then $\text{Gen}(V_R) = \text{Mod-}R$ and V_R clearly satisfies condition (d) of Theorem 2.3.8.

Conversely, a ∗-module V_R is finitely generated by Theorem 2.3.6 and is projective when $\text{Gen}(V_R) = \text{Mod-}R$ by Proposition 2.3.3.

(2) Suppose that $0 \to K \longrightarrow V^{(A)} \xrightarrow{f} M \to 0$ is exact. If V_R is a quasi-progenerator, then according to Lemma 2.4.4, $\mathrm{Gen}(V_R)$ is closed under submodules, so $\mathrm{Gen}(V_R) = \mathrm{Pres}(V_R)$. Since V is quasi-projective and finitely generated, according to [1, Proposition 16.12] Hf is epic. Thus by Theorem 2.3.8 (d), V_R, is a $*$-module. Conversely, if V_R is a $*$-module, $K \in \mathrm{Gen}(V_R)$ implies Hf epic by Theorem 2.3.8, so V_R is quasi-projective and generates its submodules when $\mathrm{Gen}(V_R)$ is closed under submodules.

(3) If V_R is a tilting module, then every injective R-module belongs to $\mathrm{Gen}(V_R)$. Since $V^{(A)} \in \mathrm{Gen}(V_R) = V^{\perp}$, an exact sequence

$$0 \to K \longrightarrow V^{(A)} \xrightarrow{f} M \to 0$$

yields an exact sequence

$$H(V^{(A)}) \xrightarrow{Hf} HM \to \mathrm{Ext}_R^1(V, K) \to 0.$$

Thus Hf is epic if and only if $K \in V^{\perp} = \mathrm{Gen}(V_R)$, so Theorem 2.3.8(e) shows that V_R is a $*$-module.

Let V_R be a $*$-module with $E(R) \in \mathrm{Gen}(V_R)$. One easily checks that $E(R_R)$ generates every injective module. From the exact sequence

$$0 \to M \longrightarrow E(M) \xrightarrow{f} E(M)/M \to 0,$$

in which both $E(M)$ and $E(M)/M$ belong to $\mathrm{Gen}(V_R)$, we obtain the exact sequence

$$0 \to HM \longrightarrow H(E(M)) \xrightarrow{Hf} H(E(M)/M) \to \mathrm{Ext}_R^1(V, M) \to 0.$$

If $M \in \mathrm{Gen}(V_R)$, then by Proposition 2.3.3 Hf is epic, and so $M \in V^{\perp}$. If $M \in V^{\perp}$, then Hf is epic, so $M \in \mathrm{Gen}(V_R)$ by Proposition 2.2.1(1). ∎

Now, after noting that by Proposition 2.3.2 any equivalence Mod-$R \approx$ Mod-S is induced by a $*$-module, we have K. Morita's seminal theorem on equivalence from 1958 [65].

Theorem 2.4.6 (Morita). *A module V_R with $S = \mathrm{End}(V_R)$ is a progenerator if and only if $H : \mathrm{Mod}\text{-}R \rightleftarrows \mathrm{Mod}\text{-}S : T$ is an equivalence.*

Proof. If V_R is a progenerator, it is a $*$-module with $\mathrm{Gen}(V_R) = \mathrm{Mod}\text{-}R$ by Theorem 2.4.5(1), and since $_S V$ is projective, $\mathrm{Cogen}(V_S^*) = \mathrm{Mod}\text{-}S$ by Proposition 2.3.5. Conversely, if $H : \mathrm{Mod}\text{-}R \rightleftarrows \mathrm{Mod}\text{-}S : T$ is an equivalence, then, according to Lemma 2.1.3, $\text{Gen}(V_R) = \text{Mod-}R$ and $\text{Cogen}(V_R^*) = \text{Mod-}S$, and V_R is a *-module. Thus Theorem 2.4.5(1) applies. ∎

Morita's theorems on equivalence are discussed in detail in [1, Chapter 6]. In particular, if V_R is a progenerator with $S = \text{End}(V_R)$, then $\text{Hom}_S(V, \_)$: $S\text{-Mod} \rightleftarrows R\text{-Mod}$: $(V \otimes_R \_)$ is a category equivalence.

Since, according to Lemma 2.4.4, $\text{Gen}(V_R)$ is closed under submodules when V is a quasi-progenerator, it follows that (as in the case of Morita equivalence) the modules corresponding under the equivalence in the next theorem have isomorphic lattices of submodules.

Theorem 2.4.7. *A module V_R with $S = \text{End}(V_R)$ is a quasi-progenerator if and only if $H : \text{Gen}(V_R) \rightleftarrows \text{Mod-}S : T$ is an equivalence.*

Proof. Assume $H : \text{Gen}(V_R) \rightleftarrows \text{Mod-}S : T$ is an equivalence. Then V_R is a *-module by Definition 2.3.1 and Lemma 2.1.3. Hence T is an exact functor by Proposition 2.3.3. We claim that $\text{Gen}(V_R)$ is closed under submodules. For, suppose that $0 \longrightarrow K \overset{f}{\longrightarrow} M \longrightarrow L \longrightarrow 0$ is exact in Mod-R with M (and hence also L) in $\text{Gen}(V_R)$. Applying H and letting $Z = \text{Coker}H(f)$, we obtain exact sequences

$$0 \to HK \overset{Hf}{\to} HM \to Z \to 0$$
$$\text{and} \qquad 0 \to TZ \overset{T(\iota)}{\to} THL$$

where $\iota : Z \to HL$ is the inclusion. Hence $\nu_L \circ T(\iota)$ is monic by Theorem 2.3.8(b). Thus we obtain the commutative diagram with exact rows

$$
\begin{array}{ccccccc}
0 \to & THK & \to THM \to & TZ & & \to 0 \\
 & \downarrow \nu_K & \downarrow \nu_M & \downarrow \nu_L \circ T(\iota) & \\
0 \to & K & \to M \to & L & & \to 0
\end{array}
$$

and conclude that ν_K is epic by the Five Lemma. Hence $K \in \text{Gen}(V_R)$ as claimed. Thus according to Theorem 2.4.5(2), V_R is a quasi-progenerator.

Conversely, if V_R is a quasi-progenerator, then V_R is a *-module with $\text{Gen}(V_R)$ closed under submodules by Theorem 2.4.5(2), and by Proposition 1.2.8, $_SV$ is flat. Hence $\text{Mod-}S = \text{Cogen}(V_S^*)$ by Proposition 2.3.5. ∎

If V is a right R module and $I = \mathbf{r}_R(V)$, the right annihilator of V in R, then we may identify Mod-R/I with $\{M \in \text{Mod-}R \mid MI = 0\}$. Thus if $S = \text{End}(V_R)$ and $_SV_{R/I}$ induces an equivalence between subcategories $\mathcal{C}$

and $\mathcal{D}$ of Mod-R/I and Mod-S, respectively, then so does $_S V_R$ viewing $\mathcal{C}$ as a subcategory of Mod-R.

Corollary 2.4.8. *If R is a right artinian ring, then V_R is a quasi-progenerator if and only if V is a progenerator in* Mod-$R/\mathbf{r}_R(V)$.

Proof. Since R is right artinian, any quasi-projective R-module is projective modulo its annihilator [1, Exercise 17.16 (2)], and $R/\mathbf{r}_R(V)$ embeds a finite direct sum of copies of any R-module V. Thus if V_R is a quasi-progenerator, then $V_{R/\mathbf{r}_R(V)}$ is a progenerator. Conversely, if $V_{R/\mathbf{r}_R(V)}$ is a progenerator, then $\mathrm{Gen}(V_R) = \mathrm{Mod}\text{-}R/\mathbf{r}_R(V)$; thus, Theorem 2.4.7 applies. ∎

Jacobson's density theorem [1, Theorem 14.4] suggests another method of obtaining examples of quasi-progenerators, as the next two results indicate.

We say that a ring B is a *ring extension* of a ring R if there is a ring homomorphism $\psi : R \to B$ so that a B-module M_B becomes an R-module M_R with scalar multiplication $mr = m\psi(r)$. Then, given V_B, we say that R (or $\psi(R)$) is *V-dense in B* if, for each $b \in B$ and each finite subset $\{v_1, \ldots, v_n\} \subseteq V$, there is an $r \in R$ such that $v_i b = v_i r$ for $i = 1, \ldots, n$.

Proposition 2.4.9. *Let V_R be a quasi-progenerator with $S = \mathrm{End}(V_R)$ and $B = \mathrm{BiEnd}(V_R) = \mathrm{End}(_S V)$. Then V_B is a quasi-progenerator and R is V-dense in B.*

Proof. Since V_R is a quasi-progenerator, $\mathrm{Gen}(V_R)$ is closed under submodules. Thus, if $X_R \leq V^n$, then $X = \mathrm{Tr}_V(X)$, and if $f : V \to X$, then there are $s_1, \ldots, s_n \in S$ such that $f(v) = (s_1 v, \ldots, s_n v)$ for all $v \in V$. But then if $b \in B$, we have $f(v)b = (s_1(vb), \ldots, s_n(vb)) = f(vb) \in X$. Thus $XB = X$ and $\mathrm{Hom}_B(V, X) = \mathrm{Hom}_R(V, X)$. In particular,

$$(v_1, \ldots, v_n)B = (v_1, \ldots, v_n)RB = (v_1, \ldots, v_n)R$$

so that R is V-dense in B. Now if $K \leq V_R$ and $g : V \to V/K$ is an R-homomorphism, let $g(v) = v' + K$. Then for each $b \in B$ there is an $r \in R$ such that $vb = vr$ and $v'b = v'r$, and recalling that $K \leq V_B$, we have

$$g(vb) = g(vr) = g(v)r = v'b + K = g(v)b.$$

Thus it follows that V_B is also a quasi-progenerator. ∎

Proposition 2.4.10. *If V_B is a quasi-progenerator with $S = \mathrm{End}(V_B)$ and R is V-dense in B, then V_R is a quasi-progenerator with $S = \mathrm{End}(V_R)$.*

Proof. Clearly density implies that R-submodules and B-submodules of V are one-and-the-same, and we have just shown in Proposition 2.4.9 that $\text{Hom}_R(V, V/K) = \text{Hom}_B(V, V/K)$ for all $K \le V$; thus, V_R is quasi-projective. Similarly, $\text{Hom}_R(V, K) = \text{Hom}_B(V, K)$, so V_R generates its submodules. ∎

Example 2.4.11. *Let S be an arbitrary ring, let $B = \mathbb{RFM}_{\mathbb{N}}(S)$, the ring of $\mathbb{N} \times \mathbb{N}$ row finite matrices over S, and let Φ be the set of matrices in B with only finitely many nonzero entries. If $e = e_{11}$ denotes the first diagonal idempotent, then eB is a (projective) quasi-progenerqator, and if R is any subring of B containing Φ, then R is eB-dense in B, and so $eB = eR$ is a quasi-progenerator over R with $\text{End}(eB) \cong S$ and $B = \text{BiEnd}(eR)$.*

In subsequent chapters we shall discuss tilting modules in detail. However, at this point we can employ part (3) of Theorem 2.4.5 to show that tilting modules and *-modules are closely related when certain finiteness conditions are satisfied by R.

Corollary 2.4.12. *If R_R is finitely cogenerated (e.g., if R is right artinian), then any faithful *-module V is a tilting module.*

Proof. By hypothesis, $R_R \hookrightarrow V^n$ for some positive integer n, so V^n, and hence V, generates every injective right R-module. ∎

If $S = \text{End}(V_R)$ and $_SV_{R/I}$ induces an equivalence between subcategories $\mathcal{C}$ and $\mathcal{D}$ of Mod-R/I and Mod-S, respectively, such that $\mathcal{C}$ is closed under direct sums and epimorphisms and $\mathcal{D}$ is closed under direct products and submodules, V_R must be a *-module. On the other hand, $MI = 0$ for all $M \in \text{Gen}(V_R)$. Thus we have

Corollary 2.4.13. *If R is a right artinian ring, then V_R is a *-module if and only if V is a tilting module in Mod-$R/\mathbf{r}_R(V)$.*

R. Colpi and G. D'Este [27] have presented examples of faithful *-modules that are neither quasi-progenerators nor tilting modules.

3

Tilting Modules

Tilting modules were originally introduced in [8], [9], and [46] as a tool in the representation (or module) theory of finite dimensional algebras. They, and more general versions of tilting modules, have found many important applications to this theory. Our concern here is their application to the module theory of more general rings.

3.1. Generalized Tilting Modules

We begin with a generalization of the notion of tilting modules that was introduced by R. Colpi and J. Trlifaj in [32].

Definition 3.1.1. A module V_R is a generalized tilting module if $\text{Gen}(V_R) = V^\perp$.

Important properties of tilting modules that hold in this more general setting are recorded in the next two propositions.

Proposition 3.1.2. *If V_R is a generalized tilting module and $S = \text{End}(V_R)$, then V_R is faithful and $_S V$ is finitely generated.*

Proof. V_R is faithful since $E(R_R) \in \text{Gen}(V_R)$, and $_S V$ is finitely generated by Lemma 1.2.3 since $V^\perp$ is closed under direct products. ∎

Proposition 3.1.3. *If V_R is a generalized tilting module, then $\text{Gen}(V_R) = \text{Pres}(V_R)$.*

Proof. Let $M_R \in \text{Gen}(V_R)$ (so $M = \text{Tr}_V(M)$) and apply Lemma 2.3.7. ∎

These propositions lead to the following characterizations of generalized tilting modules.

28

Theorem 3.1.4. *A module* $V \in$ *Mod-R is a generalized tilting module if and only if*

(i) proj. dim .$(V_R) \leq 1$;

(ii) $\text{Ext}_R^1(V, V^{(A)}) = 0$ *for all sets A;*

(iii) *there is an exact sequence* $0 \to R_R \to V_0 \to V_1 \to 0$ *where* $V_0, V_1 \in$ Add(V_R).

Proof. Assume that V_R is a generalized tilting module.

(i) follows from Proposition 1.1.1.

(ii) is obvious from the definition.

(iii) By Proposition 3.1.2 we have the exact sequence

$$0 \to R_R \to V^n \xrightarrow{f} L \to 0,$$

and by Proposition 3.1.3, Gen(V_R) = Pres(V_R); so there is an exact sequence

$$0 \to K \to V^{(A)} \xrightarrow{g} L \to 0$$

where $K \in$ Gen$(V_R) = V^\perp$. Now consider the pullback diagram

$$
\begin{array}{ccc}
0 & & 0 \\
\downarrow & & \downarrow \\
K & = & K \\
\downarrow & & \downarrow \\
0 \to R \to X & \longrightarrow & V^{(A)} \to 0 \\
\| \quad \downarrow & & g \downarrow \\
0 \to R \to V^n & \xrightarrow{f} & L \to 0 \\
\downarrow & & \downarrow \\
0 & & 0
\end{array}
$$

Since K, $V^n \in V^\perp$, we have $X \in V^\perp =$ Pres(V_R), and an exact sequence

$$0 \to N \longrightarrow V^{(B)} \longrightarrow X \to 0$$

with $N \in V^\perp$. But this sequence splits since

$$0 = \text{Ext}_R^1(V^{(A)}, N) \to \text{Ext}_R^1(X, N) \to \text{Ext}_R^1(R, N) = 0$$

is exact. Thus $X \in$ Add(V_R) and

$$0 \to R \to X \to V^{(A)} \to 0$$

verifies (iii).

Conversely, assume (i), (ii), and (iii). If $M \in \text{Gen}(V_R)$, we have an exact sequence $0 \to K \to V^{(A)} \to M \to 0$, which in turn induces the exact sequence

$$0 = \text{Ext}^1_R(V, V^{(A)}) \to \text{Ext}^1_R(V, M) \to \text{Ext}^2_R(V, K) = 0,$$

and so we have $\text{Gen}(V_R) \subseteq V^\perp$. (Note that this inclusion is implied by conditions (i) and (ii).) If $M \in V^\perp$, then the exact sequence of (iii) induces the commutative diagram

$$
\begin{array}{ccccc}
\text{Hom}_R(V_0, M) & \overset{\alpha}{\to} & \text{Hom}_R(R, M) & \cong & M \\
\uparrow \cong & & \uparrow \beta & & \\
\text{Hom}_R(V_0, \text{Tr}_{V_0}(M)) & \to & \text{Hom}_R(R, \text{Tr}_{V_0}(M)) & \cong & \text{Tr}_{V_0}(M)
\end{array}
$$

where β is monic and, since $V_1 \in \text{Add}(V)$, α is epic. It follows that β is an isomorphism, and hence $\text{Tr}_{V_0}(M) = M$. Since $V_0 \in \text{Add}(V)$, $\text{Tr}_V(M) = M$ as well. ∎

Proposition 3.1.5. *A module $V \in \text{Mod-}R$ is a generalized tilting module if and only if*

    (i) $\text{proj} . \dim .(V_R) \leq 1$;
    (ii) $\text{Ext}^1_R(V, V^{(A)}) = 0$ *for all sets A;*
    (iii) $\text{Ker}(\text{Hom}_R(V, \_)) \cap V^\perp = 0$.

Proof. If V_R is generalized tilting and $\text{Hom}_R(V, M) = 0 = \text{Ext}^1_R(V, M)$, then $M \in V^\perp = \text{Gen}(V_R)$; therefore $M = 0$, proving (iii). (i) and (ii) follow from Proposition 3.1.4.

Conversely, assume that (i), (ii), and (iii) hold. As noted in the proof of Proposition 3.1.4, (i) and (ii) imply that $\text{Gen}(V_R) \subseteq V^\perp$. So assume $M \in V^\perp$. Since $\text{Gen}(V_R) \subseteq V^\perp$, $\text{Tr}_V(M) \in V^\perp$. Hence, the exact sequence $0 \to \text{Tr}(M) \overset{\beta}{\to} M \to M/\text{Tr}(M) \to 0$, where β is the inclusion, induces an exact sequence

$$
\begin{aligned}
0 \to \text{Hom}_R(V, \text{Tr}_V(M)) & \overset{\text{Hom}(V,\beta)}{\longrightarrow} \text{Hom}_R(V, M) \\
& \to \text{Hom}_R(V, M/\text{Tr}_V(M)) \to 0
\end{aligned}
$$

where $\text{Hom}(V, \beta)$ is epic by definition of trace, and so $\text{Hom}_R(V, M/\text{Tr}(M)) = 0$. But we also have the exact sequence

$$0 = \text{Ext}^1_R(V, M) \to \text{Ext}^1_R(V, M/\text{Tr}_V(M)) \to 0;$$

thus $M/\text{Tr}(M) = 0$ by (iii), and we conclude that $M \in \text{Gen}(V_R)$. ∎

We conclude this section by showing that a generalized tilting module is a tilting module if it is self-small.

Corollary 3.1.6. *If $V \in$ Mod-R is a generalized tilting module, then V_R is a $*$-module if and only if it is self-small.*

Proof. The corollary follows from Proposition 3.1.5(ii) and Theorem 2.3.8(e). ∎

Corollary 3.1.7. *If $V \in$ Mod-R is a self-small generalized tilting module, then V_R is a tilting module.*

Proof. A self-small generalized tilting module is finitely generated by Corollary 3.1.6 and Theorem 2.3.6. ∎

3.2. Tilting Modules

The results of the preceding section yield the basic characterizations of tilting modules in the following theorem due to R. Colpi [25]. There condition (b) is the classical definition of a tilting module.

Theorem 3.2.1. *The following are equivalent for a module $V \in$ Mod-R:*

(a) V_R *is a tilting module;*
(b) V_R *is finitely presented and*
    (i) proj . dim .$(V_R) \leq 1$,
    (ii) $\mathrm{Ext}_R^1(V, V) = 0$,
    (iii) *there is an exact sequence* $0 \to R_R \to V_0 \to V_1 \to 0$ *where* $V_0, V_1 \in \mathrm{add}(V_R)$;
(c) V_R *is finitely presented and*
    (i) proj . dim .$(V_R) \leq 1$,
    (ii) $\mathrm{Ext}_R^1(V, V) = 0$,
    (iii) $\mathrm{Ker}(\mathrm{Hom}_R(V, \_)) \cap V^\perp = 0$.

Proof. $(a) \Rightarrow (b)$. Assuming (a), V_R is finitely presented by Proposition 1.1.3. Since V_R is a generalized tilting module, (b)(i) and (b)(ii) are immediate from Theorem 3.1.4. It remains to prove (b)(iii). By Proposition 3.1.2 we have, as in Proposition 1.2.5, the exact sequence

$$0 \to R_R \to V^n \to V^n/R \to 0,$$

and by Proposition 3.1.3 there is an exact sequence

$$0 \to K \to V^{(X)} \to V^n/R \to 0$$

where $K \in \mathrm{Gen}(V_R)$. Thus we obtain the exact sequence

$$\mathrm{Hom}_R(V^n, K) \overset{i_K^*}{\to} K \to \mathrm{Ext}_R^1(V^n/R, K) \to \mathrm{Ext}_R^1(V^n, K) = 0$$

where i_K^* is epic by Proposition 1.2.5; hence, we see by our hypothesis that $\mathrm{Ext}_R^1(V^n/R, K) = 0$, and so we can conclude that, being finitely generated, $V^n/R \in \mathrm{add}(V_R)$.

$(b) \Rightarrow (c)$. Since V_R is finitely presented, $\mathrm{Ext}_R^1(V, \_)$ commutes with direct sums by Proposition 1.1.2; thus, V_R is a generalized tilting module by Proposition 3.1.4. Hence, (c)(iii) follows from Proposition 3.1.5.

$(c) \Rightarrow (a)$. As in the proof of $(b) \Rightarrow (c)$, V_R is a generalized tilting module by Proposition 3.1.5. ∎

Like Morita equivalence, tilting is a two-sided concept. Indeed, like progenerators, tilting modules are faithfully balanced and are also tilting modules over their endomorphism rings.

Proposition 3.2.2. *If V_R is a tilting module with $S = \mathrm{End}(V_R)$, then ${}_S V$ is a tilting module with $R \cong \mathrm{End}({}_S V)$, canonically.*

Proof. Let

$$0 \to P_1 \longrightarrow P_0 \longrightarrow V \to 0 \tag{3.1}$$

be exact with the P_i finitely generated projective right R-modules, and let

$$0 \to R_R \longrightarrow V_0 \longrightarrow V_1 \to 0 \tag{3.2}$$

be exact with the $V_i \in \mathrm{add}(V_R)$. Since $\mathrm{Ext}_R^1(V, V) = 0$

$$0 \to \mathrm{Hom}_R(V, V) \longrightarrow \mathrm{Hom}_R(P_0, V) \longrightarrow \mathrm{Hom}_R(P_1, V) \to 0$$

is exact, and so here we have an exact sequence in S-Mod

$$0 \to {}_S S \longrightarrow V_0' \longrightarrow V_1' \to 0 \tag{3.3}$$

with the $V_i' \in \mathrm{add}({}_S V)$. Similarly, applying $\mathrm{Hom}_R(\_, V)$ to the exact sequence (3.2), we obtain an exact sequence

$$0 \to P_1' \longrightarrow P_0' \longrightarrow {}_S V \to 0 \tag{3.4}$$

with the P_i' finitely generated projective left S-modules. Since $S = \operatorname{End}(V_R)$, the evaluation map $\delta_V : V_R \to \operatorname{Hom}_S(\operatorname{Hom}_R(V, V), V)$ is an isomorphism, that is, V_R is $_SV_R$-*reflexive*. But then, according to [1, Proposition 20.13], so are V_0 and V_1. Thus, applying $\operatorname{Hom}_S(\_, V)$ to the exact sequence (3.4), we obtain a commutative diagram

$$
\begin{array}{ccccccccc}
0 & \to & R & \to & V_0 & \to & V_1 & \to & 0 \\
& & \delta_R \downarrow & & \delta_{V_0} \downarrow & & \delta_{V_1} \downarrow & & \\
0 & \to & \operatorname{Hom}_S(V, V) & \to & \operatorname{Hom}_S(P_0', V) & \to & \operatorname{Hom}_S(P_1', V) & \to & \operatorname{Ext}^1_S(V, V) \to 0
\end{array}
$$

with exact rows, in which δ_{V_0} and δ_{V_1} are isomorphisms. Now it follows that $\operatorname{Ext}^1_S(V, V) = 0$, and then that $R \cong \operatorname{End}(_SV)$, canonically. ■

In view of this last result, if V_R is a tilting module with $S = \operatorname{End}(V_R)$ we shall, on occasion, say that $_SV_R$ is a *tilting bimodule*. Note that a faithfully balanced bimodule $_SV_R$ that is finitely presented and satisfies conditions (i) and (ii) of Theorem 3.2.1 in both Mod-R and S-Mod is a tilting bimodule.

If K is a commutative artinian ring with radical $J(K)$, then (see [1, Proposition 30.6]) $\operatorname{Hom}_K(\_, E(K/J(K)))$ induces a duality on the finitely generated K-modules. A ring R is an *artin algebra* if its center K is an artinian ring and R is finitely generated as a K-module; then $D = \operatorname{Hom}_K(\_, E(K/J(K)))$ defines a duality

$$D : \operatorname{mod-}R \rightleftarrows R\operatorname{-mod} : D$$

called the *artin algebra duality*. Note, in particular, that, in this case $D(_RR_R)$ is a finitely generated two-sided injective cogenerator.

For an artin algebra R, one can restrict the definition of tilting modules to mod-R.

Proposition 3.2.3. *If R is an artin algebra, then a module V_R is a tilting module if and only if* $\operatorname{gen}(V_R) = V^\perp \cap \operatorname{mod-}R$.

Proof. One implication is obvious. So suppose that $\operatorname{gen}(V_R) = V^\perp \cap \operatorname{mod-}R$. We shall verify the conditions of Theorem 3.2.1(b). Clearly V_R is finitely presented and $\operatorname{Ext}^1_R(V, V) = 0$. If $M \in \operatorname{mod-}R$, then so is each term in the exact sequence $0 \to M \longrightarrow E(M) \longrightarrow L \to 0$. Thus we have an exact sequence

$$0 = \operatorname{Ext}^1_R(V, L) \to \operatorname{Ext}^2_R(V, M) \to \operatorname{Ext}^2_R(V, E(M)) = 0,$$

and it follows that, since R is artinian, proj . dim .$(V_R) \leq 1$. Finally, appealing directly to Lemma 2.3.7 (with X finite), rather than to Lemma 3.1.3, one can verify condition (iii) of Theorem 3.2.1(b) by imitating the proof that (a) implies (b) (iii) of Theorem 3.2.1. ∎

3.3. Tilting Torsion Theories

If V_R is a tilting module, then it follows at once from Definition 2.4.3 that $\text{Gen}(V_R)$ is a torsion class with torsion-free class $\text{Ker Hom}_R(V, \_)$. Thus it is of interest to determine which torsion theories so arise from a tilting module. Here we present an extension by Colpi and Trlifaj [32] of a characterization due to I. Assem [4] and S. Smaløf [71].

Definition 3.3.1. A torsion theory $(\mathcal{T}, \mathcal{F})$ (torsion class $\mathcal{T}$) in Mod-R is called a *tilting torsion theory (class)* if there is a tilting module V_R in Mod-R such that $\mathcal{T} = \text{Gen}(V_R)$.

For arbitrary rings we have the following necessary and sufficient conditions for a finitely generated module U_R with $\text{Gen}(U_R) \subseteq U^\perp$ (as in Proposition 1.4.4) to generate a tilting torsion class.

Proposition 3.3.2. *Suppose U_R is a finitely generated module with $\text{Gen}(U_R)$ $\subseteq U_R^\perp$. Then $\text{Gen}(U_R)$ is a tilting torsion class in Mod-R if and only if U_R is faithful and finitely generated over $\text{End}(U_R)$. In this case U_R is a tilting module if and only if $\text{Gen}(U_R) = \text{Pres}(U_R)$.*

Proof. Suppose $\text{Gen}(U_R) = \text{Gen}(V_R) = V^\perp$ where V_R is a tilting module. Since V_R is faithful and $V_R \in \text{Gen}(U_R)$, it is clear that U_R is faithful. Also, since $\text{Gen}(V_R) = \text{Gen}(U_R)$ is closed under direct products by Proposition 3.2.2 and Proposition 1.2.3, $_{\text{End}(U_R)}U$ is finitely generated by Proposition 1.2.3.

Conversely, assume that U_R is faithful, is finitely generated over both R and $\text{End}(U_R)$ and that $\text{Gen}(U_R) \subseteq U^\perp$. Referring to the discussion on page 5, assuming that $_{\text{End}(U_R)}U$ is generated by n elements, there is an exact sequence

$$0 \to R \to U^n \to U' \to 0$$

in mod-R. Then, for any M in Mod-R we have an exact sequence

$$\text{Hom}_R(U', M) \to \text{Hom}_R(U^n, M) \xrightarrow{i_M^*} M \to \text{Ext}_R^1(U', M) \to \text{Ext}_R^1(U^n, M)$$

where i_M^* is epic if and only if $M \in \text{Gen}(U_R)$ by Proposition 1.2.5. Since $\text{Gen}(U_R) \subseteq U^\perp$, it follows easily that $\text{Gen}(U_R) = U'^\perp$. Let $V_R = U_R \oplus U_R'$. Then V_R is finitely generated and $\text{Gen}(V_R) = \text{Gen}(U_R)$ since U_R' is an epimorphic image of U_R^n. Thus we have

$$V^\perp = U^\perp \cap U'^\perp = U^\perp \cap \mathrm{Gen}(U_R) = \mathrm{Gen}(U_R) = \mathrm{Gen}(V_R),$$

so V_R is a tilting module.

Now assume in addition that $\mathrm{Gen}(U_R) = \mathrm{Pres}(U_R)$. As above there is a tilting module V_R such that

$$\mathrm{Gen}(U_R) = \mathrm{Gen}(V_R) = V^\perp.$$

Also there is a, necessarily split, exact sequence

$$0 \to K \longrightarrow U^{(A)} \longrightarrow V \to 0$$

with $K \in \mathrm{Gen}(U_R)$, so $V_R \in \mathrm{add}(U_R)$. Therefore, $U^\perp \subseteq V^\perp = \mathrm{Gen}(U_R)$. ∎

For artinian rings, tilting torsion classes are characterized as follows.

Theorem 3.3.3. *Suppose $\mathcal{T}$ is a torsion class in Mod-R and R is right artinian. The following are equivalent:*

(a) *$\mathcal{T}$ is a tilting torsion class;*

(b) *$\mathcal{T}$ is closed under direct products and contains every injective module in Mod-R, and $\mathcal{T} = \mathrm{Gen}(U_R)$ for some finitely generated module U_R;*

(c) *$\mathcal{T} = \mathrm{Gen}(U_R)$ for some faithful module U_R, which is finitely generated both in Mod-R and in $\mathrm{End}(U_R)$-Mod.*

Proof. $(a) \Rightarrow (b)$. If $\mathcal{T} = \mathrm{Gen}(V_R)$ where V_R is a tilting module in Mod-R then, since $\mathcal{T} = V_R^\perp$, $\mathcal{T}$ contains all injectives and is closed under direct products.

$(b) \Rightarrow (c)$. Assume (b). Since $\mathrm{Gen}(U_R)$ is closed under products, $_{\mathrm{End}(U_R)}U$ is finitely generated by Proposition 1.2.3, and since U_R generates $E(R_R) \in \mathcal{T}$, U_R is faithful.

$(c) \Rightarrow (a)$. Assuming (c), first note that, if W_R is a direct summand of U_R such that $\mathrm{Gen}(W_R) = \mathrm{Gen}(U_R)$, then W_R is faithful since $U_R \in \mathrm{Gen}(W_R)$ and is finitely generated over its endomorphism ring by Proposition 1.2.3. Hence, if in addition $\mathrm{Gen}(W_R) \subseteq W_R^\perp$, (a) will follow from Proposition 3.3.2. Since R is right artinian, $U_R = \oplus_{i=1}^n U_i$ where each U_i is indecomposable. From the set $\{U_i\}_{i=1}^n$ choose a subset that is minimal with respect to the property that the direct sum of its members generates $\mathrm{Gen}(U_R)$. Renumbering the U_i, we can assume that this set is $\{U_i\}_{i=1}^k$ where $k \leq n$ and let $W_R = \oplus_{i=1}^k U_i$. To complete the proof it suffices to show that $\mathrm{Gen}(W_R) \subseteq W_R^\perp = U_1^\perp \cap \cdots \cap U_k^\perp$.

Let $1 \leq m \leq k$ and suppose $\mathrm{Ext}_R^1(U_m, M) \neq 0$ where $M \in \mathrm{Gen}(W_R)$. Then there is an exact sequence $0 \to M \to X \to U_m \to 0$ that is not split exact and $X \in \mathrm{Gen}(W_R)$ since $\mathrm{Gen}(W_R) = \mathcal{T}$ is a torsion class. Hence, there is an epimorphism $\pi : W_R^{(J)} \to X$ for some set J and, noting that any homomorphism $W_R^{(J)} = U_m^{(J)} \oplus Y \to U_m$ where $Y = \oplus_{i=1, i \neq m}^k U_i^{(J)}$ can be written in the form $\oplus_{j \in J} p_j \oplus q$ where each $p_j \in \mathrm{End}((U_m)_R)$ and $q \in \mathrm{Hom}_R(Y, U_m)$, we have a commutative diagram with exact rows

$$
\begin{array}{ccc}
W^{(J)} & = & U_m^{(J)} \oplus Y \\
\downarrow \pi & & \downarrow \underset{j \in J}{\oplus} p_j \oplus q \\
0 \to M \to \quad X & \overset{f}{\longrightarrow} & U_m \qquad\qquad \to 0.
\end{array}
$$

Clearly, since the given short exact sequence does not split, no p_j is an isomorphism. Hence each $p_j \in \mathrm{Rad}(\mathrm{End}((U_m)_R))$ as $\mathrm{End}((U_m)_R)$ is a local ring by [1, Lemma 12.8]. Thus

$$
U_m = f \circ \pi \left(W_R^{(J)} \right) = \sum_{j \in J} p_j(U_m) + q(Y) \subseteq \mathrm{Rad}(\mathrm{End}((U_m)_R))U_m + q(Y),
$$

so $q(Y) = U_m$ since $\mathrm{Rad}((U_m)_R))U_m$ is superfluous in U_m. But then we must have $U_m \in \mathrm{Gen}(\oplus_{i=1, i \neq m}^k U_i)$ contrary to the choice of W_R. ∎

3.4. Partial Tilting Modules

In Theorem 3.2.1, part (iii) of each characteristic condition (b) and (c) of a tilting module is generally the most difficult to verify. Thus one is led to the following notion.

Definition 3.4.1. A module U_R is a *partial tilting module* in case U_R is finitely presented and

   (i) $\mathrm{proj \, . \, dim \, .}(U_R) \leq 1$;
   (ii) $\mathrm{Ext}_R^1(U, U) = 0$.

These modules are characterized in [32] as follows.

Proposition 3.4.2. *A finitely generated module U_R is a partial tilting module if and only if $\mathrm{Gen}(U_R) \subseteq U^\perp$ and $U^\perp$ is a torsion class.*

Proof. According to Proposition 1.1.1, $\mathrm{proj \, . \, dim \, .} \, U \leq 1$ if and only if $U^\perp$ is closed under epimorphic images. Moreover, $U^\perp$ is always closed under extensions.

($\Rightarrow$) Since U is finitely presented, $U^{\perp}$ is closed under direct sums by Proposition 1.1.2. Thus $U^{\perp}$ is a torsion class whenever U is a partial tilting module. Now, if

$$0 \to K \to U^{(A)} \to M \to 0$$

is exact, we see from the exactness of

$$\text{Ext}_R^1(U, U^{(A)}) \to \text{Ext}_R^1(U, M) \to \text{Ext}_R^2(U, K)$$

that $\text{Gen}(U_R) \subseteq U^{\perp}$.

($\Leftarrow$) Since $U^{\perp}$ is a torsion class, U_R is finitely presented by Proposition 1.1.3, and $\text{Ext}_R^1(U, U) = 0$, since $U_R \in \text{Gen}(U_R) \subseteq U^{\perp}$. ∎

Analogous to generalized tilting modules, one has the following:

Definition 3.4.3. A module U_R is a *generalized partial tilting module* if $\text{Gen}(U_R) \subseteq U^{\perp}$ and $U^{\perp}$ is a torsion class.

We want to show that every generalized partial tilting module is a direct summand of a generalized tilting module. We shall employ the following lemma to do so.

Lemma 3.4.4. *Given M, $U \in \text{Mod-}R$, with $S = \text{End}(U_R)$, let $\{\varepsilon_i\}_I$ generate the right S-module $\text{Ext}_R^1(U, M)$. Then there is an exact sequence*

$$0 \to M \longrightarrow X \longrightarrow U^{(I)} \to 0$$

such that the connecting homomorphism $\partial : \text{Hom}_R(U, U^{(I)}) \to \text{Ext}_R^1(U, M)$ is an epimorphism.

Proof. Let

$$\varepsilon_i : \ 0 \to M \xrightarrow{f_i} X_i \xrightarrow{g_i} U \to 0$$

be extensions of M by U corresponding to the ε_i, and let $\sigma : M^{(I)} \to M$ with $\sigma : (m_i)_I \mapsto \sum_I m_i$. We shall show that the bottom row in the commutative diagram

$$
\begin{array}{ccccccccc}
0 \to & M^{(I)} & \xrightarrow{\oplus f_i} & \oplus_I X_i & \xrightarrow{\oplus g_i} & U^{(I)} & \to & 0 \\
 & \sigma \downarrow & & h \downarrow & & \| & & \\
0 \to & M & \xrightarrow{f} & X & \xrightarrow{g} & U^{(I)} & \to & 0,
\end{array}
$$

in which X is the pushout of $\oplus f_i$ and σ, satisfies the condition. Indeed, letting $\iota_i : U \to U^{(I)}$ and $\iota'_i : X_i \to \oplus_I X_i$ denote the canonical injections, we see that the diagram

$$
\begin{array}{ccccccccc}
0 & \to & M & \xrightarrow{f_i} & X_i & \xrightarrow{g_i} & U & \to & 0 \\
  &   & \parallel & & h\iota'_i \downarrow & & \iota_i \downarrow & & \\
0 & \to & M & \xrightarrow{f} & X & \xrightarrow{g} & U^{(I)} & \to & 0
\end{array}
$$

commutes. Thus, according to [56], Theorem 3.4, page 74, and Lemma 1.2, page 65, $\partial(\iota_i) = \varepsilon_i$ for all $i \in I$. It also follows from [56], Lemma 1.2, page 65, that ∂ is an S-homomorphism, and so the lemma is proved. ∎

As we shall see in Section 3.7, the next results will allow us to circumvent part (iii) in certain cases. The first of these is from [32].

Proposition 3.4.5. *Every generalized partial tilting module is a direct summand of a generalized tilting module.*

Proof. Let $\{\varepsilon_i\}_I$ be a set of $S = \mathrm{End}(U_R)$-generators for $\mathrm{Ext}^1_R(U, R)$. According to Lemma 3.4.4, there is an exact sequence

$$
0 \to R_R \xrightarrow{i} X \longrightarrow U^{(I)} \to 0 \tag{3.5}
$$

such that the connecting homomorphism $\partial : \mathrm{Hom}_R(U, U^{(I)}) \to \mathrm{Ext}^1_R(U, R)$ is an epimorphism. We shall show that

$$
V = U \oplus X
$$

is a generalized tilting module. From the sequence (3.5) we note that $U \in \mathrm{Gen}(X_R)$, so that

$$
\mathrm{Gen}(V_R) = \mathrm{Gen}(X_R),
$$

and we obtain an exact sequence

$$
\mathrm{Hom}_R(U, U^{(I)}) \xrightarrow{\partial} \mathrm{Ext}^1_R(U, R) \longrightarrow \mathrm{Ext}^1_R(U, X) \to 0 = \mathrm{Ext}^1_R(U, U^{(I)}).
$$

Thus, since ∂ is epic, we have

$$
X \in U^\perp, \tag{3.6}
$$

and since $U^\perp$ is a torsion class,

$$
\mathrm{Gen}(V_R) = \mathrm{Gen}(X_R) \subseteq U^\perp.
$$

If $M \in U^{\perp}$, then, applying $\mathrm{Hom}_R(\_, M)$ to the sequence (3.5), we obtain the exact sequence

$$\mathrm{Hom}_R(X, M) \xrightarrow{\mathrm{Hom}_R(i,M)} \mathrm{Hom}_R(R, M) \to \mathrm{Ext}_R^1(U^{(I)}, M)$$
$$\to \mathrm{Ext}_R^1(X, M) \to 0 = \mathrm{Ext}_R^1(R, M).$$

Thus, since $\mathrm{Ext}_R^1(U^{(I)}, M) \cong \mathrm{Ext}_R^1(U, M)^I = 0$, $\mathrm{Hom}_R(i, M)$ is epic, so $M \in \mathrm{Gen}(X_R)$ by Lemma 1.2.4, and $M \in X^{\perp}$. Therefore

$$U^{\perp} \subseteq \mathrm{Gen}(X_R) \cap X^{\perp} = \mathrm{Gen}(V_R) \cap X^{\perp}.$$

Now we have $V^{\perp} = U^{\perp} \cap X^{\perp} = U^{\perp}$ and $U^{\perp} \subseteq \mathrm{Gen}(V_R) \subseteq U^{\perp}$. Thus we have

$$\mathrm{Gen}(V_R) = V^{\perp}. \qquad \blacksquare$$

If R is an artin algebra, partial tilting modules are direct summands of tilting modules, as K. Bongartz proved in [8]. It is worth noting here that it follows easily from part (b) of Theorem 3.2.1 that whenever V_R is a tilting module over an arbitrary ring, $\mathrm{Ext}_R^1(V, R)$ is finitely generated in $\mathrm{Mod\text{-}End}(V_R)$.

Corollary 3.4.6. *If U_R is a partial tilting module such that $\mathrm{Ext}_R^1(U, R)$ is finitely generated over $S = \mathrm{End}(U_R)$ (for example, if R is an artin algebra), then U_R is a direct summand of a tilting module.*

Proof. In this case we may replace I by a positive integer n in the proof of Proposition 3.4.5. Then, since $\mathrm{Ext}_R^1(U^n, U^{(A)}) = 0$ and, by equation (3.6), $\mathrm{Ext}_R^1(U^n, X^{(A)}) = 0$, we have commutative diagrams with exact rows

$$
\begin{array}{ccccccc}
0 \to & \mathrm{Hom}_R(U^n, U)^{(A)} & \longrightarrow & \mathrm{Hom}_R(X, U)^{(A)} & \longrightarrow & \mathrm{Hom}_R(R, U)^{(A)} & \to 0 \\
& \cong\downarrow & & \downarrow & & \cong\downarrow & \\
0 \to & \mathrm{Hom}_R(U^n, U^{(A)}) & \longrightarrow & \mathrm{Hom}_R(X, U^{(A)}) & \longrightarrow & \mathrm{Hom}_R(R, U^{(A)}) & \to 0
\end{array}
$$

and

$$
\begin{array}{ccccccc}
0 \to & \mathrm{Hom}_R(U^n, X)^{(A)} & \longrightarrow & \mathrm{Hom}_R(X, X)^{(A)} & \longrightarrow & \mathrm{Hom}_R(R, X)^{(A)} & \to 0 \\
& \cong\downarrow & & \downarrow & & \cong\downarrow & \\
0 \to & \mathrm{Hom}_R(U^n, X^{(A)}) & \longrightarrow & \mathrm{Hom}_R(X, X^{(A)}) & \longrightarrow & \mathrm{Hom}_R(R, X^{(A)}) & \to 0
\end{array}
$$

from which it follows that $V_R = U \oplus X$ is self-small, so Proposition 3.1.7 applies. Of course, if R is an artin algebra, then V_R is automatically finitely presented. $\blacksquare$

3.5. The Tilting Theorem

The raison d'être of a tilting module V_R is that it (though not as thoroughly as a progenerator) provides a connection between the categories of modules over R and $S = \text{End}(V_R)$. This connection is described in the Tilting Theorem as presented in [18], earlier versions of which appeared in [9], [8], [47], [64].

Theorem 3.5.1 (Tilting Theorem). *Suppose that V_R is a tilting module in* Mod-R *with* $S = \text{End}(V_R)$. *Let*

$$H = \text{Hom}_R(V, \_), \;\; H' = \text{Ext}_R^1(V, \_), \;\; T = (\_ \otimes_S V), \;\; T' = \text{Tor}_1^S(\_, V)$$

to obtain pairs of functors

$$H : \text{Mod-}R \rightleftarrows \text{Mod-}S : T \;\; and \;\; H' : \text{Mod-}R \rightleftarrows \text{Mod-}S : T'$$

and let

$$\mathcal{T} = \text{Ker}\, H', \;\; \mathcal{F} = \text{Ker}\, H, \;\; \mathcal{S} = \text{Ker}\, T, \;\; \mathcal{E} = \text{Ker}\, T'.$$

Then

(1) $(\mathcal{T}, \mathcal{F})$ and $(\mathcal{S}, \mathcal{E})$ are torsion theories in Mod-R *and* Mod-S, *respectively;*

(2) $TH' = 0_{\text{Mod-}R} = T'H$ and $HT' = 0_{\text{Mod-}S} = H'T$;

(3) There are natural transformations θ and φ that, together with the canonical natural transformations ν and η, yield exact sequences

$$0 \to THM \xrightarrow{\nu_M} M \xrightarrow{\theta_M} T'H'M \to 0$$

and

$$0 \to H'T'N \xrightarrow{\varphi_N} N \xrightarrow{\eta_N} HTN \to 0$$

for each $M \in$ Mod-R and each $N \in$ Mod-S;

(4) The restrictions

$$H : \mathcal{T} \rightleftarrows \mathcal{E} : T \;\; and \;\; H' : \mathcal{F} \rightleftarrows \mathcal{S} : T'$$

define category equivalences.

Proof. (1) By Definition 2.4.3, $\text{Gen}(V_R) = V_R^\perp = \mathcal{T}$, and so it follows at once that, since $V_R^\perp$ is closed under extensions, $\mathcal{T} = \text{Gen}(V_R)$ is a torsion class. But then, according to Proposition 1.4.2 the corresponding torsion-free class must be $\text{Ker}\, H = \mathcal{F}$. According to Theorem 2.4.5, V_R is a ∗-module. Thus, by Proposition 2.3.5, $\text{Cogen}(V_S^*) = \text{Ker } T' = \mathcal{E}$, so $\mathcal{E}$ is a torsion-free class, and since, by adjointness,

$$\text{Hom}_S(N, V_S^*) \cong \text{Hom}_R((N \otimes_S V), C)$$

with C_R an injective cogenerator, the corresponding torsion class is $\text{Ker}(\_ \otimes_S V) = \mathcal{S}$. Thus we see that $(\mathcal{T}, \mathcal{F})$ and $(\mathcal{S}, \mathcal{E})$ are torsion theories in Mod-R and Mod-S, respectively.

(2) Since, by Lemma 2.1.3, $H : \text{Mod-}R \to \text{Cogen}(V_S^*) = \text{Ker } T'$, we have $T'H = 0_{\text{Mod-}R}$. Also $H'T = 0_{\text{Mod-}S}$, since $T : \text{Mod-}S \to \text{Gen}(V_R) = \text{Ker } H'$. Since $\mathcal{T}$ contains all injective modules in Mod-R, if $M \in \text{Mod-}R$ there is an exact sequence

$$0 \to M \longrightarrow E \longrightarrow D \to 0$$

with E injective and E, $D \in \mathcal{T}$. Thus, since V_R is a $*$-module, we obtain a commutative diagram

$$
\begin{array}{ccccc}
E & \longrightarrow & D & \longrightarrow & 0 \\
\nu_E \uparrow & & \nu_D \uparrow & & \\
THE & \longrightarrow & THD & \longrightarrow TH'M & \longrightarrow 0
\end{array}
$$

with exact rows in which the vertical maps are isomorphisms. Thus $TH' = 0_{\text{Mod-}R}$. On the other hand, since $S_S \in \mathcal{E}$, so is every submodule of a projective right S-module. Thus, given N_S, an exact sequence

$$0 \to K \longrightarrow P \longrightarrow N \to 0$$

with P projective, a similar argument shows that $HT' = 0_{\text{Mod-}S}$.

(3) Let $M \in \text{Mod-}R$ and let

$$0 \to M \xrightarrow{f} E \xrightarrow{g} D \to 0$$

be exact with E injective so that E, $D \in \mathcal{T}$. Then we have an exact sequence

$$0 \to HM \xrightarrow{Hf} HE \xrightarrow{Hg} HD \xrightarrow{\delta} H'M \to 0$$

that yields exact sequences

$$0 \to HM \xrightarrow{Hf} HE \xrightarrow{\alpha} L \to 0 \tag{3.7}$$

and

$$0 \to L \xrightarrow{\beta} HD \xrightarrow{\delta} H'M \to 0 \qquad (3.8)$$

in which β is the inclusion, $\beta \circ \alpha = Hg$ and $L = \operatorname{Im} Hg = \operatorname{Ker} \delta \in \mathcal{E} = \operatorname{Ker} T'$ by (1) and (2). Hence, from sequence (3.7) we obtain an exact sequence

$$0 \to THM \xrightarrow{THf} THE \xrightarrow{T\alpha} TL \to 0, \qquad (3.9)$$

and since $T'HD = 0 = TH'M$, from sequence (3.8) we obtain an exact sequence

$$0 \to T'H'M \xrightarrow{\partial} TL \xrightarrow{T\beta} THD \to 0 \qquad (3.10)$$

in which $\nu_D : THD \to D$ and $\nu_E : THE \to E$ are isomorphisms since V_R is a $*$-module. Thus we obtain a commutative diagram

$$
\begin{array}{ccccccc}
 & & & & & 0 & \\
 & & & & & \downarrow & \\
 & 0 & & 0 & & T'H'M & \\
 & \downarrow & & \downarrow & & \partial \downarrow & \\
0 \to & THM & \xrightarrow{THf} & THE & \xrightarrow{T\alpha} & TL & \to 0 \\
 & \nu_M \downarrow & & \nu_E \downarrow & & \nu_D \circ T\beta \downarrow & \\
0 \to & M & \xrightarrow{f} & E & \xrightarrow{g} & D & \to 0 \\
 & \pi_M \downarrow & & \downarrow & & \downarrow & \\
 & \operatorname{Coker} \nu_M & & 0 & & 0 & \\
\end{array}
$$

with exact rows and columns. Thus by the Snake Lemma there is an exact sequence

$$0 \to THM \xrightarrow{\nu_M} M \xrightarrow{\theta_M} T'H'M \to 0$$

where π_M is the canonical epimorphism, $\theta_M = s^{-1} \circ \pi_M$, and

$$s = \pi_M \circ f^{-1} \circ \nu_E \circ (T\alpha)^{-1} \circ \partial$$

is the snake, an isomorphism. Of course, if M' is another R-module and

$$0 \to M' \xrightarrow{f'} E' \xrightarrow{g'} D' \to 0$$

is exact with E' injective, there is a similar diagram

$$
\begin{array}{ccccccc}
 & & & & & 0 & \\
 & & & & & \downarrow & \\
 & 0 & & 0 & & T'H'M' & \\
 & \downarrow & & \downarrow & & \partial' \downarrow & \\
0 \to & THM' & \xrightarrow{THf'} & THE' & \xrightarrow{T\alpha'} & TL' & \to 0 \\
 & \nu_{M'} \downarrow & & \nu_{E'} \downarrow & & \nu_{D'} \circ T\beta' \downarrow & \\
0 \to & M' & \xrightarrow{f'} & E' & \xrightarrow{g'} & D' & \to 0, \\
 & \pi_{M'} \downarrow & & \downarrow & & \downarrow & \\
 & \text{Coker}\, \nu_{M'} & & 0 & & 0 &
\end{array}
$$

and here let us call the snake $t = \pi_{M'} \circ (f')^{-1} \circ \nu_{E'} \circ (T\alpha')^{-1} \circ \partial'$. Then, to establish that θ is natural, we must show that, if $h : M \to M'$, then $(t^{-1} \circ \pi_{M'}) \circ h = T'H'h \circ (s^{-1} \circ \pi_M)$, or equivalently,

$$
t \circ T'H'h \circ (s^{-1} \circ \pi_M) = \pi_{M'} \circ h.
$$

Since E' is injective, there is a commutative diagram

$$
\begin{array}{ccccccccc}
0 & \to & M & \xrightarrow{f} & E & \xrightarrow{g} & D & \to & 0 \\
 & & h \downarrow & & h' \downarrow & & h'' \downarrow & & \\
0 & \to & M' & \xrightarrow{f'} & E' & \xrightarrow{g'} & D' & \to & 0
\end{array}
\tag{3.11}
$$

with exact rows. Then, letting $k = Hh''|_L : L \to L'$, where $L = \text{Im}\, Hg$ and $L' = \text{Im}\, Hg'$, and considering the commutative diagram

$$
\begin{array}{ccccccc}
HM & \xrightarrow{Hf} & HE & \xrightarrow{Hg} & HD & \xrightarrow{\delta} & H'M \\
Hh \downarrow & & Hh' \downarrow & & Hh'' \downarrow & & H'h \downarrow \\
HM' & \xrightarrow{Hf'} & HE' & \xrightarrow{Hg'} & HD' & \xrightarrow{\delta'} & H'M'
\end{array}
$$

we have, corresponding to sequences (3.7), (3.9), and (3.10), the following three commutative diagrams with exact rows

$$
\begin{array}{ccccccccc}
0 & \to & HM & \xrightarrow{Hf} & HE & \xrightarrow{\alpha} & L & \to & 0 \\
 & & Hh \downarrow & & Hh' \downarrow & & k \downarrow & & \\
0 & \to & HM' & \xrightarrow{Hf'} & HE' & \xrightarrow{\alpha'} & L' & \to & 0,
\end{array}
\tag{3.12}
$$

$$
\begin{array}{ccccccccc}
0 & \to & THM & \xrightarrow{THf} & THE & \xrightarrow{T\alpha} & TL & \to & 0 \\
 & & THh \downarrow & & THh' \downarrow & & Tk \downarrow & & \\
0 & \to & THM' & \xrightarrow{THf'} & THE' & \xrightarrow{T\alpha'} & TL' & \to & 0
\end{array}
\tag{3.13}
$$

and

$$
\begin{array}{ccccccccc}
0 & \to & T'H'M & \xrightarrow{\partial} & TL & \xrightarrow{T\beta} & THD & \to & 0 \\
 & & T'H'h \downarrow & & Tk \downarrow & & THh'' \downarrow & & \\
0 & \to & T'H'M' & \xrightarrow{\partial'} & TL' & \xrightarrow{T\beta'} & THD' & \to & 0.
\end{array}
\tag{3.14}
$$

Now, observing that

$$
s^{-1} \circ \pi_M = \partial^{-1} \circ T\alpha \circ v_E^{-1} \circ f,
$$

we use these diagrams and the naturalness of v to calculate

$$
\begin{aligned}
t \circ T'H'h \circ (s^{-1} \circ \pi_M) &= \\
&\pi_{M'} \circ (f')^{-1} \circ v_{E'} \circ (T\alpha')^{-1} \circ (\partial' \circ T'H'h) \circ \partial^{-1} \circ T\alpha \circ v_E^{-1} \circ f \\
&= \pi_{M'} \circ (f')^{-1} \circ v_{E'} \circ (T\alpha')^{-1} \circ (Tk \circ \partial) \circ \partial^{-1} \circ T\alpha \circ v_E^{-1} \circ f \\
&= \pi_{M'} \circ (f')^{-1} \circ v_{E'} \circ (T\alpha')^{-1} \circ (Tk \circ T\alpha) \circ v_E^{-1} \circ f \\
&= \pi_{M'} \circ (f')^{-1} \circ v_{E'} \circ (T\alpha')^{-1} \circ (T\alpha' \circ THh') \circ v_E^{-1} \circ f \\
&= \pi_{M'} \circ (f')^{-1} \circ (v_{E'} \circ THh') \circ v_E^{-1} \circ f \\
&= \pi_{M'} \circ (f')^{-1} \circ (h' \circ v_E) \circ v_E^{-1} \circ f \\
&= \pi_{M'} \circ (f')^{-1} \circ (h' \circ f) \\
&= \pi_{M'} \circ (f')^{-1} \circ (f' \circ h) \\
&= \pi_{M'} \circ h
\end{aligned}
$$

as desired. A similar argument using exact sequences of the form $0 \to K \longrightarrow P \longrightarrow N \to 0$ in Mod-S with P projective yields the natural transformation $\varphi : H'T' \to 1_{\text{Mod}-S}$ with

$$
0 \to H'T'N \xrightarrow{\varphi_N} N \xrightarrow{\eta_N} HTN \to 0
$$

exact for all $N \in$ Mod-S.

(4) This follows at once from (2) and (3). ∎

Remark 3.5.2. Assuming the notation of the Tilting Theorem, we make the following observations:

1. The torsion submodules of $M \in$ Mod-R and $N \in$ Mod-S are

$$
\begin{aligned}
\tau_T(M) &= \text{Tr}_V(M) = v_M(THM)) \cong THM \quad \text{and} \\
\tau_S(N) &= \text{Ann}_N(_S V) = \text{Ker}\, \eta_N \cong H'T'M.
\end{aligned}
$$

2. Since S is closed under epimorphic images and direct sums, $\mathcal{F}$ is closed under submodules and direct products, and $T' : S \rightleftarrows \mathcal{F} : H'$ is an equivalence, letting $_R V'_S = H'(_R R_R)$, we see that V'_S is a $*$-module with $\text{End}(V'_S) \cong T'H'(R) \cong R/\tau_T(R)$ such that $T' \cong \text{Hom}_S(V', \_)$ on S and $H' \cong (\_ \otimes_R V')$ on $\mathcal{F}$. (See the discussion on page 16.)

3.6. Global Dimension and Splitting

Of course, the global dimensions of R and the endomorphism ring of an R-progenerator are the same. For a tilting module they are close. We denote the right global dimension of a ring R by r.gl.dim. (R).

Most of the results in this section were proved for artin algebras in [9], [8], [47], and [49] and extended to arbitrary rings in [18] and [64].

Proposition 3.6.1. *If $_SV_R$ is a tilting bimodule, then*

$$\text{r.gl.dim.}\,(S) \leq \text{r.gl.dim.}\,(R) + 1.$$

Proof. Let

$$\cdots \to P_3 \to P_2 \longrightarrow P_1 \longrightarrow P_0 \to N \to 0$$

be a projective resolution of $N \in \text{Mod-}S$. Then, since flat.dim. $_SV \leq 1$, by Theorem 3.2.1 and Proposition 3.2.2, the sequence

$$TP_3 \to TP_2 \longrightarrow TP_1$$

is exact. Thus, if E_R is injective, we see from the commutative diagram

$$\begin{array}{ccccc}
\text{Hom}_R(TP_1, E) & \longrightarrow & \text{Hom}_R(TP_2, E) & \longrightarrow & \text{Hom}_R(TP_3, E) \\
\cong \downarrow & & \cong \downarrow & & \cong \downarrow \\
\text{Hom}_S(P_1, HE) & \longrightarrow & \text{Hom}_S(P_2, HE) & \longrightarrow & \text{Hom}_S(P_3, HE)
\end{array}$$

that $\text{Ext}_S^2(N, HE) = 0$ so that inj . dim . $(HE) \leq 1$.

Let $M \in \text{Mod-}R$ with injective resolution

$$0 \to M \longrightarrow E_0 \xrightarrow{d_0} E_1 \xrightarrow{d_1} E_2 \xrightarrow{d_2} \cdots \xrightarrow{d_{m-1}} E_m \to 0,$$

and let $(\mathcal{T}, \mathcal{F})$ be the torsion theory in Mod-R induced by V. If $M \in \mathcal{T}$, then the sequence

$$0 \to HM \longrightarrow HE_0 \xrightarrow{Hd_0} HE_1 \xrightarrow{Hd_1} HE_2 \xrightarrow{Hd_2} \cdots \xrightarrow{Hd_{m-1}} HE_m \to 0$$

is exact by Proposition 2.3.3 since the E_i and Im d_1 all belong to $\mathcal{T}$. Thus it follows (by induction on m) that inj . dim . $HM \leq m + 1$. If $M \in \mathcal{F}$, the exact sequence

$$0 \to M \longrightarrow E_0 \longrightarrow \text{Im}\,d_0 \to 0$$

yields an exact sequence

$$HM = 0 \longrightarrow HE_0 \longrightarrow H(\text{Im}\,d_0) \to H'M \to 0 = H'E_0.$$

Thus, since $\operatorname{Im} d_0 \in \mathcal{T}$ and has injective dimension $\leq m - 1$, as we have just proved, inj . dim . $(H(\operatorname{Im} d_0)) \leq m$. Therefore, since inj . dim . $(H E_0) \leq 1$, we see from exact sequences

$$\operatorname{Ext}_S^{m+1}(N, H(\operatorname{Im} d_0)) \to \operatorname{Ext}_S^{m+1}(N, H'M) \to \operatorname{Ext}_S^{m+2}(N, H E_0),$$

for $N \in \operatorname{Mod}\text{-}S$, that inj . dim .$(H'M) \leq m$.

Now, if r.gl.dim. $R = n$, it follows from the exact sequences

$$0 \to H'T'N \xrightarrow{\varphi_N} N \xrightarrow{\eta_N} HTN \to 0$$

of the Tilting Theorem that r.gl.dim. $S \leq n + 1$. ∎

The most interesting case occurs when R is right hereditary, for then the torsion theory $(\mathcal{S}, \mathcal{E})$ on Mod-S induced by a tilting bimodule $_S V_R$ splits in the sense that each $N \in \operatorname{Mod}\text{-}S$ is a direct sum of a module in $\mathcal{S}$ and a module in $\mathcal{E}$. To prove this we shall employ

Lemma 3.6.2. *If $_S V_R$ is a tilting bimodule that induces the torsion theory $(\mathcal{T}, \mathcal{F})$ in Mod-R, then, for all L, $M \in \mathcal{T}$ and all integers $i \geq 0$*

$$\operatorname{Ext}_S^i(HL, HM) \cong \operatorname{Ext}_R^i(L, M).$$

Proof. Since $\operatorname{Gen}(V_R) = \operatorname{Pres}(V_R)$ (see Theorem 2.3.8), there is an exact sequence

$$\cdots \xrightarrow{d_3} V_2 \xrightarrow{d_2} V_1 \xrightarrow{d_1} V_0 \xrightarrow{d_0} L \to 0$$

with each $V_i \in \operatorname{Add}(V_R)$. Then each $H V_i$ is projective over S, and so, being exact by Theorem 2.3.8 since each $\operatorname{Im} d_i \in \mathcal{T}$, the sequence

$$\cdots \xrightarrow{Hd_3} H V_2 \xrightarrow{Hd_2} H V_1 \xrightarrow{Hd_1} H V_0 \xrightarrow{Hd_0} HL \to 0$$

is a projective resolution of HL. Since $H : \mathcal{T} \rightleftarrows \mathcal{E} : T$ is an equivalence, we have an isomorphism of complexes

$$
\begin{array}{ccccccccc}
0 \to & \operatorname{Hom}_R(V_0, M) & \xrightarrow{d_1^*} & \operatorname{Hom}_R(V_1, M) & \xrightarrow{d_2^*} & \operatorname{Hom}_R(V_2, M) & \xrightarrow{d_3^*} & \cdots \\
& \cong \downarrow & & \cong \downarrow & & \cong \downarrow & & \\
0 \to & \operatorname{Hom}_S(H V_0, HM) & \xrightarrow{Hd_1^*} & \operatorname{Hom}_S(H V_1, HM) & \xrightarrow{Hd_2^*} & \operatorname{Hom}_S(H V_2, HM) & \xrightarrow{Hd_3^*} & \cdots
\end{array}
$$

Thus it remains to show that $\operatorname{Ker} d_{i+1}^* / \operatorname{Im} d_i^* \cong \operatorname{Ext}_R^i(L, M)$ for all $i \geq 1$. To do so, first we note that, since $\operatorname{Ext}_R^j(V_k, M) = 0$ for all $j \geq 1$, the exact sequences

$$0 \to \operatorname{Im} d_{k+1} \xrightarrow{\beta_{k+1}} V_k \xrightarrow{\alpha_k} \operatorname{Im} d_k \to 0$$

yield isomorphisms

$$\operatorname{Ext}_R^{j-1}(\operatorname{Im} d_{k+1}, M) \cong \operatorname{Ext}_R^{j}(\operatorname{Im} d_k, M)$$

for all $j > 1$. Thus

$$\operatorname{Ext}_R^i(L, M) \cong \operatorname{Ext}_R^{i-1}(\operatorname{Im} d_1, M) \cong \cdots \cong \operatorname{Ext}_R^1(\operatorname{Im} d_{i-1}, M). \qquad (\#)$$

Now, from the commutative diagram with exact row and column

$$
\begin{array}{c}
V_{i+1} \\
d_{i+1} \downarrow \\
V_i \\
\alpha_i \downarrow \quad \overset{d_i}{\searrow} \\
0 \to \quad \operatorname{Im} d_i \quad \xrightarrow{\beta_i} V_{i-1} \xrightarrow{\alpha_{i-1}} \operatorname{Im} d_{i-1} \to 0 \\
\downarrow \\
0
\end{array}
$$

we obtain a commutative diagram with exact row and column

$$
\begin{array}{c}
\operatorname{Hom}_R(V_{i+1}, M) \\
d_{i+1}^* \uparrow \\
\operatorname{Hom}_R(V_i, M) \\
\overset{d_i^*}{\nearrow} \quad \alpha_i^* \uparrow \\
\operatorname{Hom}_R(V_{i-1}, M) \xrightarrow{\beta_i^*} \operatorname{Hom}_R(\operatorname{Im} d_i, M) \longrightarrow \operatorname{Ext}_R^1(\operatorname{Im} d_{i-1}, M) \to 0 \\
\uparrow \\
0
\end{array}
$$

from which, together with (#), it follows that

$$\operatorname{Ker} d_{i+1}^* / \operatorname{Im} d_i^* = \operatorname{Im} \alpha_i^* / \alpha_i^*(\operatorname{Im} \beta_i^*) \cong \operatorname{Ext}_R^1(\operatorname{Im} d_{i-1}, M) \cong \operatorname{Ext}_R^i(L, M)$$

as desired. ∎

Slightly more than the promised splitting is true.

Proposition 3.6.3. *Let $_S V_R$ be a tilting bimodule. The exact sequence*

$$0 \to H'T'N \xrightarrow{\varphi_N} N \xrightarrow{\eta_N} HTN \to 0$$

of the Tilting Theorem splits for all $N \in \text{Mod-}S$ if and only if $\text{Ext}_R^2(X, Y) = 0$ for all $X \in \mathcal{T}$ and all $Y \in \mathcal{F}$. In particular, $(\mathcal{S}, \mathcal{E})$ splits whenever R is right hereditary.

Proof. Let $X \in \mathcal{T}$ and $Y \in \mathcal{F}$, and let

$$0 \to Y \longrightarrow E \longrightarrow L \to 0$$

be exact with E injective, so that E and $L \in \mathcal{T}$. Then since $HY = 0$ we also have an exact sequence

$$0 \to HE \longrightarrow HL \longrightarrow H'Y \to 0.$$

Now, since by Lemma 3.6.2, $\text{Ext}_S^i(HX, HE) \cong \text{Ext}_R^i(X, E) = 0$ for $i \geq 1$, these two exact sequences yield the diagram

$$
\begin{array}{ccccc}
0 \to & \text{Ext}_R^1(X, L) & \longrightarrow & \text{Ext}_R^2(X, Y) & \to 0 \\
& \cong \downarrow & & & \\
0 \to & \text{Ext}_S^1(HX, HL) & \longrightarrow & \text{Ext}_S^1(HX, H'Y) & \to 0.
\end{array}
$$

Thus the condition is clearly sufficient, and it is necessary since, by the Tilting Theorem, any exact sequence $0 \to H'Y \to N \to HX \to 0$ must have $H'Y \cong H'T'N$ and $HX \cong HTN$. ∎

Definition 3.6.4. A ring S is *right tilted* if there is a tilting bimodule ${}_SV_R$ with R a right hereditary ring.

When S is right tilted the torsion theory $(\mathcal{S}, \mathcal{E})$ has the following properties, which also serve to determine when R is right hereditary.

Proposition 3.6.5. *Let ${}_SV_R$ be a tilting bimodule. Then R is a right hereditary ring if and only if (1) the induced torsion theory $(\mathcal{S}, \mathcal{E})$ in $\text{Mod-}S$ splits and (2) $\text{proj}\,.\,\dim\,.\,N \leq 1$ for all $N \in \mathcal{E}$. Moreover, if these conditions hold, then (3) $\text{inj}\,.\,\dim\,.\,N \leq 1$ for all $N \in \mathcal{S}$.*

Proof. ($\Rightarrow$). Assume that R is right hereditary. Then (1) holds by Proposition 3.6.3. Every projective $P_S \in \text{Mod-}S$ belongs to $\mathcal{E}$; so, if $N \in \mathcal{E}$, then, by Lemma 3.6.2 and the Tilting Theorem,

$$\text{Ext}_S^2(N, P) \cong \text{Ext}_R^2(TN, TP) = 0,$$

so, if $X \in \text{Mod-}S$, then an exact sequence

$$0 \to K \longrightarrow P \longrightarrow X \to 0,$$

together with Proposition 3.6.1, yields an exact sequence

$$0 = \text{Ext}_S^2(N, P) \longrightarrow \text{Ext}_S^2(N, X) \longrightarrow \text{Ext}_S^3(N, K) = 0$$

that establishes (2).

Moreover, we saw in the proof of Proposition 3.6.1 that if $M \in \mathcal{F} = \text{Ker } H$ in Mod-R, then inj. dim .$(H'M) \leq$ inj. dim. M. Thus (3) follows since $S = H'\mathcal{F}$.

($\Leftarrow$). Let $X \in \mathcal{T} = \text{Ker } H'$, let $M \in$ Mod-R, and consider the exact sequence

$$0 \to THM \longrightarrow M \longrightarrow T'H'M \to 0$$

with $THM \in \mathcal{T}$ and $T'H'M \in \mathcal{F}$. Here, by (2) and Lemma 3.6.2,

$$\text{Ext}_R^2(X, THM) \cong \text{Ext}_S^2(HX, HTHM) = 0,$$

and by (1) and Proposition 3.6.3, $\text{Ext}_R^2(X, T'H'M) = 0$. Thus proj. dim. $X \leq 1$ for all $X \in \mathcal{T}$.

Now suppose $M \in$ Mod-R and let

$$0 \to M \longrightarrow E \longrightarrow X \to 0$$

be exact with E injective, so that $E, X \in \mathcal{T}$. Then from the exact sequences

$$\text{Ext}_R^2(E, L) \to \text{Ext}_S^2(M, L) \to \text{Ext}_R^3(X, L)$$

with $L \in$ Mod-R, we see that proj. dim. $M \leq 1$. ∎

An artin algebra S of global dimension at most 2 with a torsion theory $(\mathcal{S}, \mathcal{E})$ in mod-S that satisfies conditions (1), (2), and (3) of Proposition 3.6.5 is called *quasi-tilted* in [45] and subsequent papers.

3.7. Grothendieck Groups

We begin this section with the definition of a group that contains useful information about mod-R for a right noetherian ring R.

Definition 3.7.1. If R is a right noetherian ring and $|$ mod-$R|$ consists of one member of each isomorphism class in mod-R, the *Grothendieck group* of mod-R is

$$K_0(\text{mod -}R) = \mathcal{A}/\mathcal{R}$$

with $\mathcal{A}$ the free abelian group with basis $|\text{mod-}R|$, and $\mathcal{R}$ is the subgroup of $\mathcal{A}$ generated by the elements of the form

$$M - K - L$$

if there is an exact sequence

$$0 \to K \longrightarrow M \longrightarrow L \to 0$$

in mod-R; so that letting $[M]$ denote the coset of M in $K_0(\text{mod} - R)$,

$$[M] = [K] + [L].$$

An inductive argument shows that, if

$$0 \to M_1 \longrightarrow M_2 \longrightarrow \cdots \longrightarrow M_{n-1} \longrightarrow M_n \to 0$$

is exact, then

$$\sum_{i=1}^{n} (-1)^i [M_i] = 0$$

in $K_0(\text{mod-}R)$. The Tilting Theorem, together with this observation and the following lemma, allows us to prove that the Grothendieck groups of a pair of right noetherian rings R and S are isomorphic if there is a tilting bimodule $_SV_R$.

Lemma 3.7.2. *If R and S are right noetherian rings and $_SV_R$ is a tilting bimodule, then*

(1) $H = \text{Hom}_R(V, M)$ and $H' = \text{Ext}_R^1(V, M)$ belong to mod-S whenever $M \in \text{mod-}R$;

(2) $T = N \otimes_S V$ and $T' = Tor_1^S(N, V)$ belong to mod-R whenever $N \in \text{mod-}S$.

Proof. (1) From the exact sequence $0 \to R_R \longrightarrow V_0 \longrightarrow V_1 \to 0$ with the $V_i \in \text{add}(V_R)$ we obtain an exact sequence

$$0 \to HR \to HV_0 \to HV_1 \to H'R \to 0$$

in mod-S since $\text{Ext}_R^1(V, V_0) = 0$ and the HV_i are finitely generated projective S-modules. If $M \in \text{mod-}R$, from an exact sequence $0 \to K \longrightarrow R^n \longrightarrow M \to 0$ we have an exact sequence

$$HR^n \to HM \to H'K \to H'R^n \to H'M \to 0$$

since proj . dim . $V \leq 1$. It follows that $H'M \in \text{mod-}S$; hence, $H'K \in \text{mod-}S$. But then, since all the other terms in the sequence belong to mod-S, so does HM.

(2) One easily checks that (2) holds for any bimodule $_SV_R$ such that R and S are right noetherian and V_R is finitely generated. ∎

Now, as promised, we can prove

Theorem 3.7.3. *If R and S are right noetherian rings and $_SV_R$ is a tilting bimodule, then, assuming the notation of the Tilting Theorem 3.5.1, there is an isomorphism*

$$\Phi : K_0(\text{mod-}R) \to K_0(\text{mod-}S)$$

such that

$$\Phi : [M] \mapsto [HM] - [H'M] \text{ and } \Phi^{-1} : [N] \mapsto [TN] - [T'N].$$

Proof. There is, by Lemma 3.7.2, a homomorphism $\overset{\wedge}{\Phi}$ from the free abelian group $\mathcal{A}$ with basis $|\text{mod-}R|$ to $K_0(\text{mod-}S)$ such that $\overset{\wedge}{\Phi} : M \mapsto [HM] - [H'M]$ for $M \in |\text{mod} -R|$. But for each generator $M - K - L$ of $\mathcal{R}$, from the corresponding exact sequence $0 \to K \longrightarrow M \longrightarrow L \to 0$, we obtain an exact sequence

$$0 \to HK \longrightarrow HM \longrightarrow HL \to H'K \longrightarrow H'M \longrightarrow H'L \to 0$$

(see Theorem 3.2.1). Thus it follows that $\mathcal{R} \subseteq \text{Ker } \overset{\wedge}{\Phi}$, and so Φ exists. On the other hand, since $_SV$ has projective (hence flat) dimension ≤ 1 by Proposition 3.2.2, if $0 \to K \longrightarrow N \longrightarrow L \to 0$ is exact in mod-S, then we see from Lemma 3.7.2 and the exact sequence

$$0 \to T'K \longrightarrow T'N \longrightarrow T'L \to TK \longrightarrow TN \longrightarrow TL \to 0$$

that there is a homomorphism $\Psi : K_0(\text{mod-}S) \to K_0(\text{mod-}R)$ such that $\Psi : [N] \mapsto [TN] - [T'N]$ for $N \in K_0(\text{mod} -S)$. Now it follows from the Tilting Theorem that, for each generator $[M]$ of $K_0(\text{mod-}R)$,

$$\begin{aligned} \Psi \circ \Phi([M]) &= \Psi([HM] - [H'M]) \\ &= [THM] - [T'HM] - [TH'M] + [T'H'M] \\ &= [M]. \end{aligned}$$

Thus $\Psi \circ \Phi = 1_{K_0(\text{mod} -R)}$, and similarly $\Phi \circ \Psi = 1_{K_0(\text{mod} -S)}$. ∎

Now suppose that R is right artinian with complete set of isomorphically distinct simple modules $X_1, \ldots, X_n$, and for each $M \in \text{mod-}R$ and each $i = 1, \ldots, n$, let $c_i(M)$ denote the number of terms in a composition series of M that are isomorphic to X_i. Then, by induction on composition length we see that if $M \in |\text{mod-}R|$, then

$$[M] = \sum_{i=1}^{n} c_i(M)[X_i]$$

in $K_0(\text{mod-}R) = \mathcal{A}/\mathcal{R}$. But also $\mathcal{R}$ is contained in the kernel the map $\mathcal{A} \to \mathbb{Z}^n$ such that $M \mapsto (c_1(M), \ldots, c_n(M))$. Thus there are epimorphisms $\mathbb{Z}^n \to K_0(\text{mod-}R)$ and $K_0(\text{mod-}R) \to \mathbb{Z}^n$ from which it follows, since they must be isomorphisms (see [1, Lemma 11.6]), that

Proposition 3.7.4. *If R is a right artinian ring with complete set of isomorphically distinct simple modules $X_1, \ldots, X_n$, then $K_0(\text{mod-}R)$ is a free abelian group with basis $[X_1], \ldots, [X_n]$.*

Now it follows that if R and S are right artinian and $_S V_R$ is a tilting bimodule, then R_R, S_S, V_R, and $_S V$ all have the same number of isomorphism classes of indecomposable direct summands.

Corollary 3.7.5. *If R and S are right artinian rings and $_S V_R$ is a tilting bimodule, then*

(1) R and S have the same number n of isomorphism classes of simple modules.

(2) There are positive integers $m_1, \ldots, m_n$ and indecomposable modules $V_1, \ldots, V_n$ such that $V_R \cong V_1^{m_1} \oplus \cdots \oplus V_n^{m_n}$.

Proof. (1) This follows at once from Theorem 3.7.3 and Proposition 3.7.4.

(2) According to (1), since the right artinian ring S has exactly n isomorphism classes of simple modules there is a complete orthogonal set of primitive idempotents e_{ij} in S with $i = 1, \ldots, n$ and, for each i, $j = 1, \ldots, m_i$ such that $e_{ij} S \cong e_{k\ell} S$ if and only if $i = k$. (See [1, Section 27].) Then since $S \cong \text{End}(V_R)$ and $\text{Hom}_R(e_{ij} V, e_{k\ell} V) \cong e_{ij} S e_{k\ell} \cong \text{Hom}_R(e_{ij} S, e_{k\ell} S)$, canonically, $V_R = \oplus_{ij} e_{ij} V$ with $e_{ij} V \cong e_{k\ell} V$ if and only if $i = k$. ∎

In practice it can be reasonably easy to determine when a given module V_R is a partial tilting module, but the third condition (iii), the existence of an exact sequence $0 \to R_R \longrightarrow V_0 \longrightarrow V_1 \to 0$ with V_0, $V_1 \in \text{add}(V_R)$, is more difficult to verify. In the artinian algebra case this problem can often be circumvented by the following result that is due to K. Bongartz [8].

Corollary 3.7.6. *Let R be a right artinian ring with n isomorphism classes of simple modules, and suppose that a partial tilting module $V_R = V_1^{m_1} \oplus \cdots \oplus V_n^{m_n}$ with the V_i pairwise non-isomorphic indecomposable modules. If the endomorphism ring of every finitely generated right R module is right artinian and $\operatorname{Ext}_R^1(V, R)$ is finitely generated over $S = \operatorname{End}(V_R)$ (for example, if R is an artin algebra), then V_R is a tilting module.*

Proof. According to Corollary 3.4.6, V_R is a direct summand of a tilting module, but then by Corollary 3.7.5 and the Krull-Schmidt theorem, V_R must be a tilting module. ∎

The following example illustrates the efficacy of this corollary.

Example 3.7.7. *Let R be an upper triangular 4×4 matrix ring over a field K so that R is a hereditary serial algebra. Let $J = J(R)$, and let $e_i = e_{ii}$ denote the ith diagonal idempotent so that the simple R-modules are $T_i = e_i R / e_i J$ for $i = 1, 2, 3, 4$. Then we can describe the structure of $R_R = e_1 R \oplus e_2 R \oplus e_3 R \oplus e_4 R$ as*

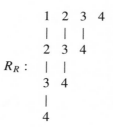

$$R_R :$$

(see [1, Page 350]). Now, letting

$$V_R = e_1 R \oplus R/Re_3 R,$$

we see that $V = V_1 \oplus V_2 \oplus V_3 \oplus V_4$ has the following structure:

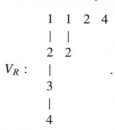

$$V_R :$$

Of course, V_R is finitely presented, and (i) proj . dim .$(V_R) \le 1$ since R is hereditary. Since V_1 is both injective and projective and $V_2 \oplus V_3 \oplus V_4$ is *projective over R modulo, the idempotent ideal Re_3R, it easily follows that (ii) $\mathrm{Ext}^1_R(V, V) = 0$. Thus V_R is a partial tilting module. But, although neither condition (iii) of Theorem 3.2.1 is immediately evident, since V is a direct sum of four isomorphically distinct indecomposable modules, V_R is a tilting module by Corollary 3.7.6. Also, calculating $S = \mathrm{End}(V_R)$ as the algebra of 4×4 matrices $[\gamma_{ij}]$ with $\gamma_{ij} \in \mathrm{Hom}_R(V_j, V_i)$, we find that*

$$
S \cong
\begin{bmatrix}
K & 0 & 0 & K \\
K & K & K & K \\
0 & 0 & K & 0 \\
0 & 0 & 0 & K
\end{bmatrix}
\Big/ Ke_{24}
$$

so that the structure of S_S is indicated by

$$
S_S : \quad
\begin{array}{cccc}
1 & 2 & 3 & 4 \\
| & \swarrow \searrow & & \\
4 & 1 \quad 3 & &
\end{array}
\quad .
$$

Thus gl.dm. $S = 2$. Also, since Mod-R consists of the direct sums of copies of the factors of the e_iR (see [1, Theorem 32.3)]), it follows that

$$
\mathcal{T} = \mathrm{Add}(e_1R \oplus e_1R/e_1J^3 \oplus e_1R/e_1J^2 \oplus e_1R/e_1J \oplus e_2R/e_2J \oplus e_4R)
$$

and

$$
\mathcal{F} = \mathrm{Ker}\, H = \mathrm{Add}(e_2R \oplus e_3R).
$$

So, since $(\mathcal{S}, \mathcal{E})$ splits by Proposition 3.6.3, S has exactly eight indecomposable right modules.

The finitely generated projective modules over any ring yield another type of Grothendieck group that we shall use to obtain a version of Corollaries 3.7.5 and 3.7.6 for semiperfect noetherian rings of finite global dimension.

Definition 3.7.8. If R is a ring and $|\mathrm{proj}$-$R|$ consists of one member of each isomorphism class of finitely generated projective modules, then the *Grothendieck group of* proj-R

$$
K_0(\mathrm{proj}\text{-}R) = \mathcal{A}_p/\mathcal{R}_p
$$

with $\mathcal{A}_p$ the free abelian group on $|\mathrm{proj}$-$R|$ and $\mathcal{R}_p$ is the subgroup of $\mathcal{A}_p$ generated by the elements of the form

$$
P - P' - P''
$$

if there is a (split) exact sequence

$$0 \to P' \longrightarrow P \longrightarrow P'' \to 0$$

in proj-R; so that letting $[P]_p$ denote the equivalence class of P in $K_0(\text{proj-}R)$,

$$[P]_p = [P']_p + [P'']_p.$$

If R is right noetherian, then there is clearly an epimorphism $\varphi : [P]_p \mapsto [P]$ of $K_0(\text{proj-}R)$ onto the subgroup $\mathcal{P}$ of $K_0(\text{mod-}R)$ generated by the equivalence classes of projective modules in $|\text{mod-}R|$. If moreover $M \in \text{mod-}R$ has a finite projective resolution

$$0 \to P_m \longrightarrow \cdots \longrightarrow P_1 \longrightarrow P_0 \longrightarrow M \to 0,$$

then

$$[M] = \sum_{i=0}^{m} (-1)^i [P_i] \in \mathcal{P}.$$

Thus if the right global dimension of R is finite, $\varphi : K_0(\text{proj-}R) \to K_0(\text{mod-}R)$ is an epimorphism. In fact, using arguments common to homological algebra, it can be shown as in [73, Theorem 4.4, page 102] that in this case φ is an isomorphism with inverse $[M] \mapsto \sum_{i=0}^{m}(-1)^i[P_i]_p$.

If R is semiperfect with complete irredundant set $P_1, \ldots, P_n$ of indecomposable right projective modules (see [1, Section 27]), then, since every finitely generated projective right R-module is of the form $P_1^{m_1} \oplus \cdots P_n^{m_n}$, we see similarly to the discussion proceeding Proposition 3.7.4, that

Proposition 3.7.9. *If R is a semiperfect ring with complete set of isomorphically distinct indecomposable projective right modules $P_1, \ldots, P_n$, then $K_0(\text{proj-}R)$ is a free abelian group with basis $[P_1]_p, \ldots, [P_n]_p$.*

Since the simple right modules over the semiperfect ring R of Proposition 3.7.9 are just $P_1/P_1J, \ldots, P_n/P_nJ$, where J is the radical of R, that proposition and the two paragraphs preceding it now yield

Corollary 3.7.10. *If R and S are semiperfect right noetherian rings of finite right global dimension and $_SV_R$ is a tilting bimodule, then*

(1) R and S have the same number n of isomorphism classes of simple modules.

(2) There are positive integers $m_1, \ldots, m_n$ and pairwise non-isomorphic indecomposable modules $V_1, \ldots, V_n$ such that $V_R \cong V_1^{m_1} \oplus \cdots \oplus V_n^{m_n}$.

Corollary 3.7.11. *Let R be a semiperfect right noetherian ring of finite right global dimension such that the endomorphism ring of every finitely generated right R-module is a semiperfect right noetherian ring, and suppose that R has n isomorphism classes of simple modules. Let $V_R = V_1^{m_1} \oplus \cdots \oplus V_n^{m_n}$ be a partial tilting module with the V_i pairwise non-isomorphic indecomposable modules. If $\operatorname{Ext}_R^1(V, R)$ is finitely generated over $S = \operatorname{End}(V_R)$, then V_R is a tilting module.*

Proof. According to Corollary 3.4.6, V_R is a direct summand of a tilting module, say $W_R = V \oplus X$ is tilting. But then by Corollary 3.7.10 and the Azumaya–Krull–Schmidt theorem [1, Theorem 12.6], V_R must be a tilting module. ■

Any noetherian serial ring of finite global dimension satisfies the hypotheses of Corollary 3.7.11 according to Propositions B.1.6 and B.1.7. The following is a case in point:

Example 3.7.12. *Let D be a local noetherian ring with maximal ideal M such that every one-sided ideal of D is a power of M (for example let $D = K[[x]]$, the ring of power series over a field K with $\mathrm{M} = Dx$), and let R be the 4×4 $[D : \mathrm{M}]$ upper triangular matrix ring*

$$R = \begin{bmatrix} D & D & D & D \\ \mathrm{M} & D & D & D \\ \mathrm{M} & \mathrm{M} & D & D \\ \mathrm{M} & \mathrm{M} & \mathrm{M} & D \end{bmatrix}.$$

Then, letting e_1, e_2, e_3, e_4 denote the diagonal idempotents, the display

$$
\begin{array}{cccc}
1 & 2 & 3 & 4 \\
| & | & | & | \\
2 & 3 & 4 & 1 \\
| & | & | & | \\
3 & 4 & 1 & 2 \\
| & | & | & | \\
4 & 1 & 2 & 3 \\
| & | & | & | \\
1 & 2 & 3 & 4 \\
| & | & | & | \\
2 & 3 & 4 & 1 \\
\vdots & \vdots & \vdots & \vdots
\end{array}
$$

R_R :

describes the structure of R_R. Now letting

$$V_R = R/Re_3R \oplus e_4R,$$

we see that $V = V_1 \oplus V_2 \oplus V_3 \oplus V_4$ has the following structure:

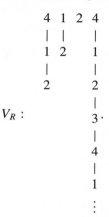

$$V_R :$$

Then V_1, V_2, V_3 are all projective over R/Re_3R so, $\mathrm{Ext}_R^1(V_i, V_j) = 0$ for $1 \le i, j \le 3$ and $\mathrm{Ext}_R^1(V_4, V_i) = 0$ for $1 \le i \le 4$ since $V_4 = e_4R$. Since $e_4R \cong e_3J$,

$$0 \to V_4 \longrightarrow E(V_4) \longrightarrow E(e_3R/e_3J) \to 0$$

is the minimal injective resolution of V_4 (see Section B.2). Thus we have epimorphisms

$$\mathrm{Hom}_R(V_i, E(e_3R/e_3J)) \longrightarrow \mathrm{Ext}_R^1(V_i, V_4) \to 0.$$

But for $i = 1, 2, 3$, no V_i has a composition factor isomorphic to e_3R/e_3J, and so $\mathrm{Hom}_R(V_i, E(e_3R/e_3J)) = 0$. Thus we finally see that (ii) $\mathrm{Ext}_R^1(V, V) = 0$. Since R is hereditary by Corollary B.1.4, we have (i) proj. dim .$(V_R) = 0$. Also, by Propositions B.1.2 and B.1.6, R satisfies the hypotheses of Corollary 3.7.11, and so V_R is a tilting module.

A complete classification of the tilting modules over noetherian serial rings and their endomorphism rings was given in [18].

3.8. Torsion Theory Counter Equivalence

The Tilting Theorem (3.5.1) gives rise to torsion theories $(\mathcal{T}, \mathcal{F})$ and $(\mathcal{S}, \mathcal{E})$ in Mod-R and Mod-S, respectively, with category equivalences $\mathcal{T} \approx \mathcal{E}$ and

$\mathcal{S} \approx \mathcal{F}$. As noted in Proposition 2.3.2 and in Remark 3.5.2, these equivalences are representable by certain $*$-modules. Here we shall discuss a more general notion that was introduced in [19].

Definition 3.8.1. Let $(\mathcal{T}, \mathcal{F})$ and $(\mathcal{S}, \mathcal{E})$ be torsion theories in Mod-R and Mod-S, respectively. A pair of category equivalences

$$F : \mathcal{T} \rightleftarrows \mathcal{E} : G \quad \text{and} \quad F' : \mathcal{F} \rightleftarrows \mathcal{S} : G'$$

is called a *torsion theory counter equivalence*.

We shall show that both equivalences in a torsion theory counter equivalence are induced by representable functors

$$H : \text{Mod-}R \rightleftarrows \text{Mod-}S : T \quad \text{and} \quad H' : \text{Mod-}R \rightleftarrows \text{Mod-}S : T'$$

that satisfy orthogonality relations and admit exact sequences like those in (2) and (3) of the Tilting Theorem, and, conversely, that any two pairs of functors satisfying such orthogonality relations and admitting such exact sequences induce a torsion theory counter equivalence.

Theorem 3.8.2. *Let $(\mathcal{T}, \mathcal{F})$ and $(\mathcal{S}, \mathcal{E})$ be torsion theories in* Mod-R *and* Mod-S, *respectively, and suppose that*

$$F : \mathcal{T} \rightleftarrows \mathcal{E} : G \quad \text{and} \quad F' : \mathcal{F} \rightleftarrows \mathcal{S} : G'$$

is a torsion theory counter equivalence. Let

$$_S V_R = G(S/\tau_S(S)) \quad \text{and} \quad _R V'_S = F'(R/\tau_T(R))$$

be the induced bimodules and let

$$H = Hom_R(V, \_), \ T = (V \otimes_S \_), \ H' = (\_ \otimes_R V'), \ and \ T' = Hom_S(V', \_).$$

Then

(1) $F \cong H|_\mathcal{T}$, $G \cong T|_\mathcal{E}$, $F' \cong H'|_\mathcal{F}$, *and* $G' \cong T'|_\mathcal{S}$;

(2) $TH' = 0_{\text{Mod-}R} = T'H$ *and* $HT' = 0_{\text{Mod-}S} = H'T$;

(3) *The canonical natural transformations induce exact sequences*

$$0 \to THM \xrightarrow{\nu_M} M \xrightarrow{\theta_M} T'H'M \to 0$$

and

$$0 \to H'T'N \xrightarrow{\phi_N} N \xrightarrow{\eta_N} HTN \to 0$$

with

$$\nu_M(THM) = \tau_T(M) \quad and \quad \phi_N(H'T'N) = \tau_S(N)$$

for all $M \in$ Mod-R and all $N \in$ Mod -S.

Proof. By Proposition 1.4.3

$$\tau_S(S) = \cap\{\mathbf{r}_S(N)|N \in \mathcal{E}\} = \mathbf{r}_S(\mathcal{E})$$

and

$$\tau_T(R) = \cap\{\mathbf{r}_R(M)|M \in \mathcal{F}\} = \mathbf{r}_R(\mathcal{F}).$$

Let $\bar{S} = S/\tau_S(S)$ and $\bar{R} = R/\tau_T(R)$. By Proposition 2.3.2, (1) is true, V_R is a $*$-module with $\bar{S} = \text{End}(V_R)$, and V'_S is a $*$-module with $\bar{R} = \text{End}(V'_S)$. Moreover, $\mathcal{T} = \text{Gen}(V_R)$, $\mathcal{E} = \text{Cogen}(V_{\bar{S}}^*)$, $\mathcal{S} = \text{Gen}(V'_S)$, and $\mathcal{F} = \text{Cogen}(V_{\bar{R}}'^*)$ (where $V_{\bar{S}}^* = \text{Hom}_R(_{\bar{S}}V_R, C_R)$ and $V_{\bar{R}}'^* = \text{Hom}_S(_{\bar{R}}V'_S, C'_S)$ for injective cogenerators C_R and C'_S in Mod-R and Mod-S, respectively). In particular, it follows that $HM \in \mathcal{E}$ and $T'N \in \mathcal{F}$ for all $M \in$ Mod-R and $N \in$ Mod-S. Now, $V_R \in \mathcal{T}$ and so, since $T'N \in \mathcal{F}$ for any $N \in$ Mod-S,

$$\begin{aligned}\text{Hom}_S((V \otimes_R V'), N) &\cong \text{Hom}_R(V, \text{Hom}_S(V', N), \\ &= \text{Hom}_R(V, T'N) \\ &= 0.\end{aligned}$$

Thus

$$HT'N = \text{Hom}_R(V, T'N) = 0$$

and $V \otimes_R V' = 0$, so

$$H'TN = (N \otimes_S V) \otimes_R V' = 0.$$

This and a similar argument establish (2). Let ν and η be the canonical natural transformations associated with H and T and let ϕ and θ be the canonical natural transformations associated with T' and H'. As noted on page 16, we may assume that the given equivalences are given by the restrictions of these natural transformations. Then $\text{Im } \nu = \text{Ker } \theta$ and $\text{Im } \phi = \text{Ker } \eta$ by Proposition 1.4.3. Also, ν_M is monic for all $M \in$ Mod-R and θ_M is epic for all $M \in$ Mod-$\bar{R}$ by Theorem 2.3.8. To verify that θ_M is epic for all $M \in$ Mod-R, employing (2) and (1), we see that the exact sequence

$$0 \to \tau_T(M) \overset{i}{\to} M \overset{n}{\to} M/\tau_T(M) \to 0$$

yields an isomorphism

$$0 \to T'H'M \stackrel{T'H'n}{\to} T'H'(M/\tau_T(M)) \to 0.$$

Thus, since $T'H'n \circ \theta_M = \theta_{M/\tau_T(M)} \circ n$ is epic, it follows that θ_M is epic, so

$$0 \to THM \stackrel{\nu_M}{\to} M \stackrel{\theta_M}{\to} T'H'M \to 0$$

is exact, and similarly,

$$0 \to H'T'N \stackrel{\phi_N}{\to} N \stackrel{\eta_N}{\to} HTN \to 0$$

is too. ∎

Orthogonality relations and exact sequences like those of Theorem 3.8.2 induce torsion theories and counter equivalences between them.

Theorem 3.8.3. *Suppose that there are pairs of functors*

$$H : \text{Mod-}R \rightleftarrows \text{Mod-}S : T \quad and \quad H' : \text{Mod-}R \rightleftarrows \text{Mod-}S : T'$$

such that

$$TH' = 0_{\text{Mod-}R} = T'H \quad and \quad HT' = 0_{\text{Mod-}S} = H'T$$

and that there are natural transformations ν, θ, ϕ, η that induce exact sequences

$$0 \to THM \stackrel{\nu_M}{\to} M \stackrel{\theta_M}{\to} T'H'M \to 0 \text{ and } 0 \to H'T'N \stackrel{\phi_N}{\to} N \stackrel{\eta_N}{\to} HTN \to 0$$

for all $M \in \text{Mod-}R$ and all $N \in \text{Mod-}S$. Let

$$\mathcal{T} = \text{Ker } H', \quad \mathcal{F} = \text{Ker } H, \quad \mathcal{S} = \text{Ker } T, \quad \mathcal{E} = \text{Ker } T',$$

and let

$$_RV_S = T(_SS_S) \quad and \quad _SV'_R = H'(_RR_R)$$

be the canonically induced bimodules. Then

(1) $(\mathcal{T}, \mathcal{F})$ and $(\mathcal{S}, \mathcal{E})$ are torsion theories in R-Mod and S-Mod, respectively.

(2) The restrictions

$$H : \mathcal{T} \rightleftarrows \mathcal{E} : T \quad and \quad H' : \mathcal{F} \rightleftarrows \mathcal{S} : T'$$

are category equivalences.

(3) $H \cong \mathrm{Hom}_R(V, \_)$, $T \cong (V \otimes_S \_)$, $H' \cong V' \otimes_R \_$, *and* $T' \cong \mathrm{Hom}_S (V', \_)$.

Proof. First note that the orthogonality relations and the exact sequences of the hypothesis establish that $HM \in \mathcal{E}$, $TN \in \mathcal{T}$, $H'M \in \mathcal{S}$, $T'N \in \mathcal{F}$ for all $M \in \mathrm{Mod}\text{-}R$ and all $N \in \mathrm{Mod}\text{-}S$. Also they imply ν_M is an isomorphism if and only if $M \in \mathcal{T}$, η_N is an isomorphism if and only if $N \in \mathcal{E}$, θ_M is an isomorphism if and only if $M \in \mathcal{F}$, and ϕ_N is an isomorphism if and only if $N \in \mathcal{S}$. Thus the equivalences of (2) are verified. For (1), suppose $\mathrm{Hom}_R(M, F) = 0$ for all $F \in \mathcal{F}$. Then $\theta_M = 0$, since $T'H'M \in \mathcal{F}$. Hence ν_M is an isomorphism so $M \in \mathcal{T}$. Conversely, if $M \in \mathcal{T}$, $F \in \mathcal{F}$, and $f \in \mathrm{Hom}_R(M, F)$, then θ_F is an isomorphism, $\theta_M = 0$, and $\theta_F \circ f = T'H'(f) \circ \theta_M = 0$ so $f = 0$. Thus we have verified condition (1) of Definition 1.4.1. Condition (2) of Definition 1.4.1 is also easily verified, and we conclude that $(\mathcal{T}, \mathcal{F})$ and similarly $(\mathcal{S}, \mathcal{E})$ are torsion theories with torsion submodules given by

$$\tau_{\mathcal{T}}(M) = \nu_M(THM) \quad \text{and} \quad \tau_S(N) = \phi_N(H'T'N)$$

for all $M \in \mathrm{Mod}\text{-}R$ and $N \in \mathrm{Mod}\text{-}S$. To establish (3) we first observe that

$$TS \cong THTS \cong T(S/\tau_S(S))$$

and that, by Proposition 1.4.3,

$$\tau_S(S) = \cap\{\mathbf{r}_S(N)|N \in \mathcal{E}\} = \mathbf{r}_S(\mathcal{E}).$$

Thus $\mathcal{E}$ is a subcategory of $\mathrm{Mod}\text{-}S/\tau_S(S)$, and letting

$$_S V_R = T(_S S_S),$$

we have $H \cong \mathrm{Hom}_R(V, \_)$ on $\mathcal{T}$ and $T \cong \_ \otimes_S V$ on $\mathcal{E}$ as in Proposition 2.3.2. Now, if $M \in \mathrm{Mod}\text{-}R$, then, since $HM \in \mathcal{E}$,

$$HM \cong HTHM \cong \mathrm{Hom}_R(V, THM) \cong \mathrm{Hom}_R(V, \mathrm{Tr}_V(M)) \cong \mathrm{Hom}_R(V, M).$$

Similarly, letting

$$V' = H'(_R R_R) = H'(R/\tau_V(R)),$$

we obtain $T' \cong \mathrm{Hom}_S(V', \_)$ on $\mathcal{S}$ and $H' \cong \_ \otimes_S V$ on $\mathcal{F}$. Moreover, with the notations of Definition 2.3.1 we have $\mathcal{F} = \mathrm{Gen}(V'_S)$ and $\mathcal{E} = \mathrm{Cogen}(V_S^*)$, and hence $\mathrm{Hom}_R(V' \otimes_S V_R, C_R) \cong \mathrm{Hom}_S(V'_S, V_S^*) = 0$. Thus $_R V' \otimes_S V = 0$, and hence for any $N \in \mathrm{Mod}\text{-}S$ we have $H'T'N \otimes_S V = T'N \otimes_R V' \otimes_S V = 0$. Thus $TN \cong THTN \cong HTN \otimes_S V \cong N \otimes_S V$. Similarly, $H' \cong \_ \otimes_R V'$ on $\mathrm{Mod}\text{-}R$. ∎

In view of the last two theorems, if H, T, H', T' are functors

$$H : \text{Mod-}R \rightleftarrows \text{Mod-}S : T \quad \text{and} \quad H' : \text{Mod-}R \rightleftarrows \text{Mod-}S : T'$$

such that

$$TH' = 0_{\text{Mod-}R} = T'H \quad \text{and} \quad HT' = 0_{\text{Mod-}S} = H'T,$$

and ν, θ, ϕ, η are natural transformations that induce exact sequences

$$0 \to THM \overset{\nu_M}{\to} M \overset{\theta_M}{\to} T'H'M \to 0 \text{ and } 0 \to H'T'N \overset{\varphi_N}{\to} N \overset{\eta_N}{\to} HTN \to 0$$

for all $M \in \text{Mod-}R$ and all $N \in \text{Mod-}S$, we shall let

$$_S V_R = T(_S S_S) \quad \text{and} \quad _R V'_S = H'(_R R_R)$$

be the canonically induced bimodules, identify

$$H = \text{Hom}_R(V, \_), T = (V \otimes_S \_), H' = (\_ \otimes_R V'), \quad \text{and} \quad T' = \text{Hom}_S(V', \_),$$

identify ν, θ, ϕ, η with the canonical natural transformations, and say that the pairs of functors H, T and H', T', or the bimodules $_S V_R$ and $_R V'_S$, *induce a torsion theory counter equivalence* between Mod-R and Mod-S, or specifically, between $(\mathcal{T}, \mathcal{E})$ and $(\mathcal{S}, \mathcal{F})$ with $\mathcal{T} = \text{Ker } H'$, $\mathcal{F} = \text{Ker } H$, $\mathcal{S} = \text{Ker } T$, and $\mathcal{E} = \text{Ker } T'$.

As an immediate consequence of Theorems 3.8.3 and 3.8.2, we have

Corollary 3.8.4. *Bimodules* $_S V_R$ *and* $_R V'_S$ *induce a torsion theory counter equivalence if and only if, for all* $M \in \text{Mod-}R$ *and all* $N \in \text{Mod-}S$,

(1) ν_M and ϕ_N are monic and θ_M and η_N are epic; and
(2) $\text{Tr}_V(M) = \text{Ann}_M(V')$ and $\text{Tr}_{V'}(N) = \text{Ann}_N(V)$.

We have a seemingly more easily verified characterization in

Theorem 3.8.5. *Bimodules* $_S V_R$ *and* $_R V'_S$ *induce a torsion theory counter equivalence if and only if the following four conditions hold:*

(1) V_R and V'_S are finitely generated, $\text{Gen}(V_R) = \text{Pres}(V_R)$, $\text{Gen}(V'_S) = \text{Pres}(V'_S)$, $\text{End}(V_R) \cong S/\ell_S(V)$, and $\text{End}(V'_S) \cong R/\ell_R(V')$, canonically;
(2) $V \otimes_R V' = 0$ and $V' \otimes_S V = 0$;
(3) $\text{Gen}(V_R) \subseteq V_R^\perp$ and $\text{Gen}(V'_S) \subseteq V'^\perp_S$;
(4) $\text{Tr}_V(M) = 0$ implies $\text{Ann}_M(V') = 0$, and $\text{Tr}_{V'}(N) = 0$ implies $\text{Ann}_N(V) = 0$ for all for all $M \in \text{Mod-}R$, $N \in \text{Mod-}S$.

Proof. ($\Rightarrow$). Note that $\ell_S(V) = \mathbf{r}_S(V_S^*) = \text{Rej}_{Cogen(V_S^*)}(S) = \text{Tr}_S(S)$ by Proposition 1.4.3 and similarly $\ell_R(V') = \text{Tr}_T(R)$. Since V_R and V_S' are *-modules, (1) follows from Theorems 2.3.6 and 2.3.8. That conditions (2) and (4) are necessary follows from the statements and proofs of Theorems 3.8.3 and 3.8.2 and Corollary 3.8.4. Since $\text{Gen}(V_R)$ is a torsion class, (1) and Proposition 2.3.3 imply (3) since H preserves the exactness of any short exact sequence $0 \to M \to M' \to V \to 0$ with M, and hence also M', in $\text{Gen}(V_R)$, yielding a splitting of the given sequence.

($\Leftarrow$). Note that by Theorem 2.3.8(d), conditions (1) and (3) imply that V_R and V_S' are *-modules. Let $\mathcal{T} = \text{Gen}(V_R)$ and $\mathcal{S} = \text{Gen}(V_S')$. Then it follows easily from (3) (see Proposition 1.4.4) that $\mathcal{T}$ and $\mathcal{S}$ are closed under extensions. Thus $\mathcal{T}$ and $\mathcal{S}$ are torsion classes, so if H, H', T, T' are induced by $_SV_R$ and $_RV_S'$ as above and

$$\mathcal{F}_0 = \text{Ker } H \quad \text{and} \quad \mathcal{E}_0 = \text{Ker } T',$$

then $(\mathcal{T}, \mathcal{F}_o)$ and $(\mathcal{S}, \mathcal{E}_o)$ are torsion theories. Let C_R be an injective cogenerator, $V_S^* = H(C)$ and $\mathcal{E} = \text{Cogen}(V_S^*)$, so that $\ell_S(V) = \mathbf{r}_S(V_S^*)$. Then V_R is a *-module with $S/\ell_S(V) \cong \text{End}(V_R)$ and $H : \mathcal{T} \rightleftarrows \mathcal{E} : T$ is an equivalence. Since

$$\text{Hom}_S(V', V^*) \cong \text{Hom}_R((V' \otimes_S V), C) = 0,$$

we see that $\mathcal{E} \subseteq \mathcal{E}_0$, and since $\text{Ann}_N(_SV) = \text{Rej}_{V_S^*}(N)$, we have $\mathcal{E}_0 \subseteq \mathcal{E}$ by (4). Thus $\mathcal{E} = \mathcal{E}_0$ and similarly $\mathcal{F} = \mathcal{F}_0$. Thus our desired conclusion follows from Theorem 3.8.2. ∎

In view of Theorem 3.8.5(3), Proposition 3.3.2 and Theorem 2.3.8(c) yield the following corollary.

Corollary 3.8.6. *Suppose* $_SV_R$ *and* $_RV_S'$ *induce a torsion theory counter equivalence. If* V_R *is faithful and* $_SV$ *is finitely generated, then* V_R *is a tilting module.*

If S is a ring with subring A and ideal I such that $S = A \oplus I$, we say that S is a *split extension* of A by I, and we write $S = A \propto I$. If, in addition, $I^2 = 0$, then $S = A \propto I$ is called a *trivial extension* of A by I. One can use these notions and the following corollary to find non-tilting examples of torsion theory counter equivalences

Corollary 3.8.7. *Let* $_AV_R$ *and* $_RV_A'$ *induce a torsion theory counter equivalence. If* $S = A \propto I$ *with* $I \otimes_A V = 0$ *and* $V' \otimes_A I = 0$, *then* $_SV_R$ *and* $_RV_S'$ *induce a torsion theory counter equivalence.*

Proof. Clearly we need only verify the "S" part of conditions (1)–(4) of Theorem 3.8.5.

Condition (1) holds because V_S' and every module it generates or cogenerates is annihilated by I, and (2) follows because $V' \otimes_S V = V' \otimes_A V$. If $N \in \text{Gen}(V_S')$, and $0 \to N \longrightarrow X \longrightarrow V' \to 0$ in Mod-S, then $N \otimes_A I = 0 = V' \otimes_A I$ implies $X \otimes_A I = 0$, so $XI = 0$. Thus the sequence splits in both Mod-A and Mod-S, and therefore (3) is verified. Finally, observing that $_S V \cong S \otimes_A V$ and $V_S' \cong V' \otimes_A S$, we have

$$
\begin{aligned}
\text{Hom}_S(V', N) &\cong \text{Hom}_S((V' \otimes_A S), N) \\
&\cong \text{Hom}_A(V', \text{Hom}_S(S, N)) \\
&\cong \text{Hom}_A(V', N))
\end{aligned}
$$

and

$$
\begin{aligned}
N \otimes_S V &\cong N \otimes_S (S \otimes_A V) \\
&\cong (N \otimes_S S) \otimes_A V \\
&\cong N \otimes_A V
\end{aligned}
$$

so that, for all $N \in \text{Mod-}S$, $\text{Tr}_{V'}(N) = 0$ implies $\text{Ann}_N(V) = 0$. ■

Example 3.8.8. *Suppose that R is a hereditary artin algebra with duality $D = \text{Hom}_K(\_, E(K/J(K)))$, and let $_R U_R = D(R)$. Then it follows easily from Theorem 3.2.1 that $_R U_R$ is a tilting bimodule. Suppose further that R has no non-zero injective projective modules, and let S denote the trivial extension*

$$
S = R \propto U = \left\{ \begin{bmatrix} r & u \\ 0 & r \end{bmatrix} \mid r \in R, \ u \in U \right\}.
$$

Then we may view U as an ideal of S and let $_S V_R$ be the bimodule with underlying set U and $UV = 0$. Let $\mathcal{T} = \text{Gen}(V_R)$ and $\mathcal{F} = \text{Ker Hom}_R(V, \_)$, so that $(\mathcal{T}, \mathcal{F})$ is a tilting torsion theory, and note that

$$
\mathcal{T} = \{ M_R \mid M \text{ is injective} \}
$$

and

$$
\mathcal{F} = \{ M_R \mid M \text{ has no injective direct summands} \}.
$$

Let $_R V_S' = \text{Ext}_R^1(V, R)$. Then by the Tilting Theorem $V' \otimes_R U = TH'(R) = 0$. Moreover $D(U \otimes_R V) \cong \text{Hom}_R(U, DV) \cong \text{Hom}_R(U, R) = 0$, since R has no non-zero injective projective modules. Thus by Corollary 3.8.7, $_S V_R$ and $_R V_S'$ induce a torsion theory counter equivalence. Now

$$
V_S^* = \text{Hom}_R(_S V_R, D(R)) = R_S \cong S/U_S
$$

so that, since V_R is a ∗-module with $\operatorname{End}(V_R) \cong S/U$,

$$\mathcal{E} = \operatorname{Cogen}(V_S^*) = \{N_S \mid N_R \text{ is projective}\}$$

is a torsion-free class in Mod-S. *Moreover, noting that if* $N \in$ Mod-S *and* $N_R = N' \oplus P$ *with* P_R *projective, then* $NU \subseteq N'$, *one easily checks that the corresponding torsion class in* Mod-S *is*

$$\mathcal{S} = \{N_S \mid N_R \text{ has no projective direct summands}\}.$$

4

Representable Dualities

Suppose that $\mathcal{A}_R$ and $_S\mathcal{B}$ are subcategories of Mod-R and S-Mod, respectively. A *duality* between $\mathcal{A}_R$ and $_S\mathcal{B}$ is a pair of contravariant functors

$$D_A : \mathcal{A}_R \rightleftarrows {}_S\mathcal{B} : D_B$$

such that $D_B \circ D_A \cong 1_{\mathcal{A}_R}$ and $D_A \circ D_B \cong 1_{_S\mathcal{B}}$. As pointed out in [1, Lemma 23.4], D_A and D_B are right adjoints of one another in the sense that there are isomorphisms

$$\mu_{MN} : \operatorname{Hom}_R(M, D_A(N)) \to \operatorname{Hom}_S(N, D_B(M))$$

that are natural in $M \in \mathcal{A}_R$ and $N \in {}_S\mathcal{B}$.

If $_SU_R$ is a bimodule we let $\Delta_{U_R} = \operatorname{Hom}_R(\_, U)$ and $\Delta_{_SU} = \operatorname{Hom}_S(\_, U)$ to obtain a pair of contravariant functors

$$\Delta_{U_R} : \operatorname{Mod-}R \rightleftarrows S\text{-}\operatorname{Mod} : \Delta_{_SU}$$

(both of which will, when convenient, simply be denoted by Δ_U or Δ). We say that a duality $D_A : \mathcal{A}_R \rightleftarrows {}_S\mathcal{B} : D_B$ is *representable by* $_SU_R$ if there are natural isomorphisms

$$D_A \cong \Delta_{U_R}|\mathcal{A}_R \quad \text{and} \quad D_B \cong \Delta_{_SU}|_S\mathcal{B}.$$

When this is the case, we make the identifications

$$D_A = \Delta_{U_R} = \Delta \quad \text{and} \quad D_B = \Delta_{_SU} = \Delta$$

and write the duality as

$$\Delta : \mathcal{A}_R \rightleftarrows {}_S\mathcal{B} : \Delta.$$

4.1. The U-dual

Throughout this section we assume that $U = {}_S U_R$ is a given bimodule and that all modules are either right R-modules or left S-modules. The functors

$$\Delta_{U_R} : \text{Mod-}R \rightleftarrows S\text{-}\text{Mod} : \Delta_{SU}$$

are adjoint on the right via

$$\mu_{MN} : \text{Hom}_R(M, \Delta_{SU}(N)) \to \text{Hom}_S(N, \Delta_{U_R}(M))$$

with

$$\mu_{MN}(f)(n) : m \mapsto f(m)(n).$$

(In fact, it is not difficult to show that any pair of functors $D : \text{Mod-}R \rightleftarrows S\text{-Mod} : D'$ that are adjoint on the right are representable by the $S - R$-bimodule $D({}_R R_R) \cong D'({}_S S_S)$.) Associated with this adjunction are the evaluation maps

$$\delta_X : X \to \Delta^2 X \text{ with } \delta_X(x) : f \mapsto f(x)$$

for X in Mod-R or S-Mod, $x \in X$, and $f \in \Delta X$. They satisfy

$$\Delta(\delta_X) \circ \delta_{\Delta X} = 1_{\Delta X} \tag{4.1}$$

and yield natural transformations

$$\delta : 1_{\text{Mod-}R} \to \Delta^2 \text{ and } \delta : 1_{S\text{-Mod}} \to \Delta^2.$$

A module X is $(U\text{-})$*reflexive* if δ_X is an isomorphism, and $(U\text{-})$*torsionless* if δ_X is a monomorphism. Note that

$$\text{Ker}\,\delta_X = \text{Rej}_U(X),$$

so that X is U-torsionless if and only if X is cogenerated by U. On the other hand, we have

Proposition 4.1.1. *If X is a module, then δ_X is an epimorphism if and only if $X/\text{Rej}_U(X)$ is U-reflexive.*

Proof. Let $X \xrightarrow{p} X/\text{Rej}_U(X)$ be the canonical epimorphism. Then by definition of the reject, $\Delta(p)$ is an isomorphism. Thus $\Delta^2(p)$ is also an isomorphism and the commutativity of the diagram

$$\begin{array}{ccc}
X & \xrightarrow{p} & X/\text{Rej}_U(X) \quad \to 0 \\
\delta_X \downarrow & & \delta_{X/\text{Rej}_U(X)} \downarrow \\
\Delta^2 X & \xrightarrow{\Delta^2 p} & \Delta^2(X/\text{Rej}_U(X))
\end{array}$$

completes the proof. ∎

Dual to Theorem A.2.2 one obtains

Proposition 4.1.2. *If $_SU_R$ induces a representable duality $\Delta : \mathcal{A}_R \rightleftarrows {_S\mathcal{B}} : \Delta$ between subcategories of* Mod-R *and* S-Mod, *then the modules in* $\mathcal{A}_R$ *and* $_S\mathcal{B}$ *are U-reflexive.*

The equation (4.1) yields

Proposition 4.1.3. *For all X in* Mod-R *or* S-Mod,

(1) $\Delta(\delta_X)$ is a split epimorphism and $\delta_{\Delta X}$ is a split monomorphism;
(2) ΔX is U-torsionless;
(3) If X is U-reflexive, then so is ΔX.

Thus, in particular, the Δ_U's induce a duality between the categories of $_SU_R$-reflexive modules in Mod-R and S-Mod.

If $M \in$ Mod-R, then, applying Δ to an exact sequence

$$R^{(B)} \longrightarrow R^{(A)} \longrightarrow M \to 0,$$

we obtain an exact sequence

$$0 \to \Delta M \longrightarrow U^A \longrightarrow U^B$$

in S-Mod so that $\Delta M \in \text{Copres}(_SU)$. If M_R is finitely presented, we may assume that A and B are finite, and then we say that ΔM is *finitely copresented* by $_SU$. If $M \in$ mod-R, so that A may be finite, we say that ΔM is *semi-finitely copresented* by $_SU$. Thus we have

Proposition 4.1.4. *For all X in* Mod-R *or* S-Mod, *and a bimodule $_SU_R$,*

(1) $\Delta X \in \text{Copres}(U)$;
(2) If X is finitely presented, then ΔX is finitely copresented by U;
(3) If X is finitely generated, then ΔX is semi-finitely copresented by U.

Note that, since $\text{Ker}\,\delta_R = \text{Rej}_U(R) = \mathbf{r}_R(U)$ and $\Delta^2(_SS) \cong \text{End}(U_R)$, we have

Proposition 4.1.5. *Relative to a bimodule $_SU_R$,*

(1) U_R is faithful if and only if R_R is U-torsionless;
(2) $S \cong \text{End}(U_R)$, canonically, if and only if $_SS$ (and hence U_R) is U-reflexive.

4.2. Costar Modules

As we just observed in Proposition 4.1.5, if $S = \mathrm{End}(U_R)$, then both U_R and $_S S$ are U-reflexive. A module U_R with $S = \mathrm{End}(U_R)$ is called a *costar module* if every U-torsionless right R-module whose U-dual is finitely generated is U-reflexive and every finitely generated torsionless left S-module is U-reflexive. Thus a module U_R with $S = \mathrm{End}(U_R)$ is a costar module if $_S U_R$ induces a duality

$$\Delta_U : \mathcal{D}_R \rightleftarrows {_S}\mathcal{C} : \Delta_U$$

where

$$\mathcal{D}_R = \{M_R \in \mathrm{Cogen}(U_R) \mid \Delta_U M \in S\text{-}\mathrm{mod}\} \text{ and } {_S}\mathcal{C} = S\text{-}\mathrm{mod} \cap \mathrm{Cogen}(_S U).$$

Here, and in later sections, we shall investigate the extent to which the role of costar modules is dual to that of $*$-modules.

Throughout the section we shall assume that $S = \mathrm{End}(U_R)$ and $\mathcal{D}_R$ and $_S\mathcal{C}$ are the categories of modules whose objects are those described in this last display. Our goal is to determine when $_S U_R$ induces a duality between them.

We begin with the following lemma, dual to Lemma 2.3.7, in order to obtain a characterization of $\mathcal{D}_R$.

Lemma 4.2.1. *Given a bimodule $_S U_R$, suppose $M \in \mathrm{Mod}\text{-}R$ and the left S-module ΔM is generated by elements $\{\gamma_a \mid a \in A\}$. Then there is a monomorphism $M / \mathrm{Rej}_U(M) \xrightarrow{f} U^A$ such that Δf is an epimorphism. Thus if $C = \mathrm{Coker}\, f$, there are exact sequences*

$$0 \to \Delta C \longrightarrow \Delta(U^A) \xrightarrow{\Delta f} \Delta(M / \mathrm{Rej}_U(M)) \to 0$$

and

$$0 \to \mathrm{Ext}_R^1(C, U) \longrightarrow \mathrm{Ext}_R^1(U^A, U) \xrightarrow{\mathrm{Ext}_R^1(f, U)} \mathrm{Ext}_R^1(M / \mathrm{Rej}_U(M), U).$$

Proof. Letting $\pi_a : U^A \to U$ denote the canonical projections for the direct product U^A, define $\varphi : M \to U^A$ via $\pi_a(\varphi(m)) = \gamma_a(m)$. Then $\mathrm{Ker}\, \varphi = \mathrm{Rej}_U(M)$ and φ induces the desired monomorphism f since $\Delta f(\pi_a) = \pi_a \circ f = \gamma_a$ for all $a \in A$. ∎

Now we have

Proposition 4.2.2. *A module $M \in \mathcal{D}_R$ if and only if there is a positive integer n and monomorphism $M \xrightarrow{f} U^n$ such that Δf is an epimorphism. In particular, $\mathcal{D}_R \subseteq \mathrm{cogen}(U_R)$.*

Proof. The condition is necessary by Lemma 4.2.1. Conversely, the condition yields an epimorphism $S^n \cong \Delta U^n \xrightarrow{\Delta f} \Delta M$. ∎

As suggested earlier, we intend to prove dual versions of results in Chapter 2. To do so we need dual versions of Lemmas 2.2.1 and 2.2.2.

Lemma 4.2.3. *Suppose* $0 \to M \xrightarrow{f} X \to L \to 0$ *is exact where* X *is* U-*reflexive and* Δf *is an epimorphism. Then* M *is* U-*reflexive if and only if* L *is* U-*torsionless.*

Proof. Applying Δ we obtain the exact sequence $0 \to \Delta L \to \Delta X \xrightarrow{\Delta f} \Delta M \to 0$ and then the commutative diagram with exact rows

$$
\begin{array}{ccccccccc}
0 & \to & M & \xrightarrow{f} & X & \to & L & \to & 0 \\
 & & \delta_M \downarrow & & \delta_X \downarrow & & \delta_L \downarrow & & \\
0 & \to & \Delta^2 M & \xrightarrow{\Delta^2 f} & \Delta^2 X & \to & \Delta^2 L. & &
\end{array}
$$

Since δ_X is an isomorphism, the lemma follows from the Snake Lemma. ∎

Lemma 4.2.4. *Let* $0 \to M \xrightarrow{f} X \longrightarrow L \to 0$ *be exact in* Mod-R. *If* X *is* U-*reflexive and* L *is* U-*torsionless, then* Δf *is epic if and only if* $\mathrm{Im}\,\Delta f$ *is* U-*reflexive.*

Proof. From the exact sequence

$$0 \to \Delta L \longrightarrow \Delta X \xrightarrow{\pi} \mathrm{Im}\,\Delta f \to 0$$

and the inclusion

$$0 \to \mathrm{Im}\,\Delta f \xrightarrow{j} \Delta M$$

we obtain the commutative diagram with exact rows

$$
\begin{array}{ccccccccc}
0 & \to & M & \xrightarrow{f} & X & \longrightarrow & L & \to & 0 \\
 & & \alpha \downarrow & & \delta_X \downarrow & & \delta_L \downarrow & & \\
0 & \to & \Delta\,\mathrm{Im}\,\Delta f & \xrightarrow{\Delta\pi} & \Delta^2 X & \to & \Delta^2 L & &
\end{array}
$$

where $\alpha = \Delta j \circ \delta_M$. Since X is reflexive and L is torsionless, α is an isomorphism by the Five Lemma, so $\Delta\alpha$ is also an isomorphism. By the adjointness of the Δ functors we have

$$\Delta(\delta_M) \circ \delta_{\Delta M} = 1_{\Delta M},$$

and since δ is a natural transformation,

$$\delta_{\Delta M} \circ j = \Delta^2 j \circ \delta_{\mathrm{Im}\,\Delta f}.$$

Also $\Delta\alpha = \Delta(\Delta j \circ \delta_M) = \Delta(\delta_M) \circ \Delta^2(j)$, so

$$j = 1_{\Delta M} \circ j = \Delta(\delta_M) \circ \delta_{\Delta M} \circ j = \Delta(\delta_M) \circ \Delta^2 j \circ \delta_{\mathrm{Im}\,\Delta f} = \Delta\alpha \circ \delta_{\mathrm{Im}\,\Delta f}.$$

Hence Δf is epic if and only if j is an isomorphism if and only if $\delta_{\mathrm{Im}\,\Delta f}$ is an isomorphism. ∎

For convenience we shall denote the category of modules semi-finitely copresented by U_R as scopres(U_R). Thus, according to Proposition 4.1.4, $\Delta_U : {}_S\mathcal{C} \to$ scopres(U_R) for an arbitrary bimodule ${}_S U_R$.

One may view the modules satisfying the conditions of the next proposition as dual to weak $*$-modules.

Proposition 4.2.5. *Assume $S = End(U_R)$. The following are equivalent:*

(a) $\Delta_U :$ scopres$(U_R) \rightleftarrows {}_S\mathcal{C} : \Delta_U$ is a duality;
(b) Every finitely generated U-torsionless left S-module is U-reflexive;
(c) If $0 \to M \xrightarrow{f} U^n \to L \to 0$ is exact with $L \in$ Cogen(U_R), then $\Delta(f)$ is an epimorphism.

Proof. $(a) \Rightarrow (b)$. This is obvious, in view of Proposition 4.1.2.

$(b) \Rightarrow (c)$. Since $\mathrm{Im}\,\Delta f$ is an epimorph of ${}_S S^n$ and is contained in ΔM whenever $0 \to M \xrightarrow{f} U^n$ is exact, this implication follows from Lemma 4.2.4.

$(c) \Rightarrow (a)$. By (c), it follows from Proposition 4.2.2 that scopres$(U_R) \subseteq \mathcal{D}_R$, so $\Delta :$ scopres$(U_R) \to {}_S\mathcal{C}$. Thus we have $\Delta :$ scopres$(U_R) \rightleftarrows {}_S\mathcal{C} : \Delta$. If $M \in$ scopres(U_R), then, applying (c) and Lemma 4.2.3, we conclude that M is reflexive. Suppose ${}_S N \in {}_S\mathcal{C}$. Then there is an exact sequence $0 \to K \to {}_S S^m \xrightarrow{g} N \to 0$ and, applying Δ, we obtain an exact sequence $0 \to \Delta(N) \xrightarrow{\Delta g} \Delta(S^m) \to L \to 0$ where $\Delta(S^m) \cong U^m$ and $L \le \Delta K \in$ Cogen(U_R). Again by (c), $\Delta^2 g$ is epic, and we obtain that N is reflexive from the commutative diagram

$$\begin{array}{ccccc} S^m & \xrightarrow{\ g\ } & N & \to 0 \\ \delta_S \downarrow & & \delta_N \downarrow & \\ \Delta^2 S^m & \xrightarrow{\Delta^2 g} & \Delta^2 N & \to 0 \end{array}$$

with exact rows. ∎

Any injective module U_R satisfies condition (c) of Proposition 4.2.5. More generally

Example 4.2.6. A module U_R with $S = \text{End}(U_R)$ induces a duality Δ_U : $\text{scopres}(U_R) \rightleftarrows {}_S\mathcal{C} : \Delta_U$ whenever $\text{Cogen}(U_R) \subseteq \text{Ker Ext}^1_R(\_, U)$.

The next theorem serves to characterize costar modules with conditions that are dual to those of Theorem 2.3.8.

Theorem 4.2.7. *Assume that* $S = \text{End}(U_R)$,

$$\mathcal{D}_R = \{M_R \in \text{Cogen}(U_R) \mid \Delta_U M \in S\text{-mod}\} \text{ and } {}_S\mathcal{C} = S\text{-mod} \cap \text{Cogen}({}_SU).$$

Then the following are equivalent:

- *(a)* $\Delta : \mathcal{D}_R \rightleftarrows {}_S\mathcal{C} : \Delta$ *is a duality. That is,* U_R *is a costar module;*
- *(b)* $\Delta : \text{scopres}(U_R) \rightleftarrows {}_S\mathcal{C} : \Delta$ *is a duality and* $\mathcal{D}_R = \text{scopres}(U_R)$;
- *(c)* δ_M *is an epimorphism if* $\Delta M \in S\text{-mod}$, *and* δ_N *is an epimorphism if* $N \in S\text{-mod}$;
- *(d)* $\mathcal{D}_R \subseteq \text{scopres}(U_R)$ *and, if* $0 \to M \xrightarrow{f} U^n \to L \to 0$ *is exact with* $L \in \text{Cogen}(U_R)$, *then* $\Delta(f)$ *is an epimorphism;*
- *(e) If* $0 \to M \xrightarrow{f} U^n \to L \to 0$ *is exact in* $\text{Mod-}R$, *then* $L \in \text{Cogen}$ (U_R) *if and only if* $\Delta(f)$ *is an epimorphism.*

Proof. $(a) \Leftrightarrow (b)$. One implication is obvious. Assuming (a), $\mathcal{D}_R = \Delta({}_S\mathcal{C}) \subseteq$ $\text{scopres}(U_R)$ by Proposition 4.1.4. But If $M \in \text{scopres}(U_R)$ via an exact sequence $0 \to M \xrightarrow{f} U^n \to L \to 0$ with $L \in \text{Cogen}(U_R)$, then $\text{im} \Delta f \in {}_S\mathcal{C}$ is U-reflexive by (a), so $M \in \mathcal{D}_R$ by Lemma 4.2.4 and Proposition 4.2.2.

$(b) \Leftrightarrow (d)$. This equivalence is by Propositions 4.2.5 and 4.2.2.

$(a) \Rightarrow (c)$. Lemma 4.1.1 applies.

$(c) \Rightarrow (a)$. This will follow from Lemma 4.1.1 if we show that $\Delta :$ ${}_S\mathcal{C} \to \mathcal{D}_R$. As in the proof that (a) implies (b), we see that $\text{scopres}(U_R) \subseteq \mathcal{D}_R$. Now in view of Proposition 4.1.4, we do have $\Delta : {}_S\mathcal{C} \to \mathcal{D}_R$.

$(d) \Rightarrow (e)$. Suppose $0 \to M \xrightarrow{f} U^n \to L \to 0$ is exact with Δf epic. Then $M \in \mathcal{D}_R$ by Proposition 4.2.2, so M is reflexive since (d) implies (a). Since U^n is reflexive by hypothesis, $L \in \text{Cogen}(U_R)$ by Lemma 4.2.3.

$(e) \Rightarrow (b)$. By Proposition 4.2.5 it suffices to show that $\mathcal{D}_R = \text{scopres}(U_R)$. If $M \in \mathcal{D}_R$, then by Proposition 4.2.2 there is an exact sequence $0 \to M \xrightarrow{f}$ $U^n \to L \to 0$ with Δf epic. Hence (e) implies that $M \in \text{scopres}(U_R)$. If $M \in \text{scopres}(U_R)$, then (e) also implies $M \in \mathcal{D}_R$ by Proposition 4.2.2. ∎

A costar module actually satisfies a condition that is a little stronger than condition (c) of Theorem 4.2.7.

Proposition 4.2.8. *Let U_R be a costar module inducing the duality Δ_U : $\mathcal{D}_R \rightleftarrows {}_S\mathcal{C} : \Delta_U$. If $0 \to K \xrightarrow{f} M \longrightarrow L \to 0$ is exact with $M \in \mathcal{D}_R$, then Δf is epic if and only if L is U-torsionless. In this case, $K \in \mathcal{D}_R$.*

Proof. Assume $M \in \mathcal{D}_R$ and $0 \to K \xrightarrow{f} M \longrightarrow L \to 0$ is exact. First suppose that Δf is epic. Then $\Delta M \in {}_S\mathcal{C}$, so ΔK is finitely generated. Thus by Theorem 4.2.7(c), δ_K is epic and hence is an isomorphism (and also $K \in \mathcal{D}_R$). Thus $L \in \mathrm{Cogen}(U_R)$ by Lemma 4.2.3. Conversely, assume that $L \in \mathrm{Cogen}(U_R)$. By hypothesis ΔM is finitely generated, and so is $\mathrm{Im}\,\Delta f \subseteq \Delta K$. But then $\mathrm{Im}\,\Delta f$ is U-reflexive because it belongs to ${}_S\mathcal{C}$. Thus Δf is epic by Lemma 4.2.4. ∎

Corollary 4.2.9. *If U_R is a costar module inducing the duality Δ_U : $\mathcal{D}_R \rightleftarrows {}_S\mathcal{C} : \Delta_U$, then the Δ's preserve exactness of short exact sequences of modules in both $\mathcal{D}_R$ and ${}_S\mathcal{C}$.*

Proof. By Proposition 4.2.8, Δ_U preserves exactness of sequences of modules in $\mathcal{D}_R$.

If $0 \to K \xrightarrow{g} N \xrightarrow{f} L \to 0$ is exact with K, N, L in ${}_S\mathcal{C}$, then in the exact sequence

$$0 \to \Delta L \xrightarrow{\Delta f} \Delta N \xrightarrow{\Delta g} \Delta K$$

$\Delta N \in \mathcal{D}_R$ is U-reflexive and $\mathrm{Im}\,\Delta g$ is U-torsionless. Thus the bottom row is exact in the commutative diagram

$$
\begin{array}{ccccccccc}
0 \to & K & \xrightarrow{g} & N & \xrightarrow{f} & L & \to 0 \\
& \downarrow & & \delta_N \downarrow & & \delta_L \downarrow & \\
0 \to & \Delta(\mathrm{Im}\,\Delta g) \to & & \Delta^2 N & \xrightarrow{\Delta^2 f} & \Delta^2 L & \to 0,
\end{array}
$$

so, since N and L are U-reflexive, $\Delta(\mathrm{Im}\,\Delta g) \cong K$ is finitely generated and $\mathrm{Im}\,\Delta g \in \mathcal{D}_R$ is U-reflexive. Thus Δg is epic by Lemma 4.2.4. ∎

Now, modulo the next lemma, we are in position to show that over an artin algebra, a finitely generated costar module is just the dual of a ∗-module.

Lemma 4.2.10. *If R is an artin algebra and $U_R \in \mathrm{mod}\text{-}R$, then $\mathcal{D}_R = \mathrm{cogen}(U_R)$ and ${}_S\mathcal{C} = \mathrm{cogen}({}_S U)$.*

Proof. Since R is an artin algebra, so is $S = \mathrm{End}(U_R)$, and a module is finitely generated if and only if it is finitely cogenerated. Thus $\mathcal{C}_S = \mathrm{cogen}(U_S)$. Also $\mathrm{cogen}(U_R) \subseteq \mathcal{D}_R$ since $\mathrm{Hom}_R(M, U) \subseteq \mathrm{Hom}_K(M, U)$ where K is the center of R, and $\mathcal{D}_R \subseteq \mathrm{cogen}(U_R)$ by Proposition 4.2.2. ∎

Proposition 4.2.11. *Let R be an artin algebra and let D : mod-$R \rightleftarrows R$-mod : D denote the artin algebra dual. Then a finitely generated module U_R is a costar module if and only if $_R V = D(U)$ is a $*$-module.*

Proof. ($\Rightarrow$). From Lemma 4.2.10, assuming that U_R is a costar module with $S = \mathrm{End}(U_R)$, it follows that $\Delta_{U_R} : \mathrm{cogen}(U_R) \rightleftarrows \mathrm{cogen}(_S U) : \Delta_{_S U}$ is a duality. According to Corollary 2.4.13, we may assume that $_R V$ is faithful and prove that it is a tilting module. Now U_R is also faithful, so that $R_R \in \mathrm{cogen}(U_R)$. Thus $D(R_R) \in \mathrm{gen}(_R V) = D(\mathrm{cogen}(U_R))$ and so is every finitely generated injective left R-module. Let $H = \Delta_{U_R} \circ D : R\text{-mod} \to S\text{-mod}$, and observe that

$$H(M) = \mathrm{Hom}_R(DM, U) \cong \mathrm{Hom}_R(V, M),$$

naturally. Let $M \in R$-mod to obtain an exact sequence $0 \to M \xrightarrow{f} E(M) \xrightarrow{g} L \to 0$ from which we obtain an exact sequence

$$H(E(M)) \xrightarrow{Hg} H(L) \to \mathrm{Ext}_R^1(V, M) \to 0.$$

But

$$0 \to D(L) \xrightarrow{Dg} D(E(M)) \xrightarrow{Df} D(M) \to 0$$

is exact, and if $M \in \mathrm{gen}(_R V)$, all three modules in this sequence belong to $\mathrm{cogen}(U_R)$ so that, by Corollary 4.2.9, $Hg = \Delta_{U_R}(Dg)$ is epic. Thus we see that $\mathrm{gen}(_R V) \subseteq \mathrm{Ker}\,\mathrm{Ext}_R^1(V, \_)$. On the other hand, suppose that $\mathrm{Ext}_R^1(V, M) = 0$. Then $Hg = \Delta_{U_R}(Dg)$ is epic, so by Proposition 4.2.8, $D(M) \in \mathrm{cogen}(U_R)$ and hence $M \in \mathrm{gen}(_R V)$. Thus the implication follows from Proposition 3.2.3.

($\Leftarrow$). Suppose that $_R V$ is a $*$-module with $S = \mathrm{End}(_R V)$ and let $_S U_R = D(_R V_S)$. Then

$$_S V_R^* = \mathrm{Hom}_R(V, D(R)) \cong \mathrm{Hom}_R(R, D(V)) \cong {}_S U_R.$$

Thus letting $H = \mathrm{Hom}_R(V, \_)$ and $T = (\_ \otimes_S V)$, we see from Definition 2.3.1 that

$$H \circ D : \mathrm{cogen}(U_R) \rightleftarrows \mathrm{cogen}(_S U) : D \circ T$$

is a duality. But $H \circ D \cong \Delta_{U_R}$ and $D \circ T \cong \Delta_{_S U}$. ∎

Now applying Lemma 4.2.10, Proposition 4.2.11, and the artin algebra dual again, one can easily obtain

Corollary 4.2.12. *A finitely generated module $_R V$ with $S = \operatorname{End}(_R V)$ over an artin algebra R is a $*$-module if and only if*

$$\operatorname{Hom}_R(V, \_) : \operatorname{gen}(_R V) \rightleftarrows \operatorname{cogen}(_S D(V)) : V \otimes_S \_$$

is an equivalence of categories.

4.3. Quasi-Duality Modules

A subcategory $\mathcal{C}$ of Mod-R or S-Mod is *finitely closed* if all submodules, epimorphic images, and finite direct sums of modules in $\mathcal{C}$ belong to $\mathcal{C}$.

Definition 4.3.1. A module U_R with $S = \operatorname{End}(U_R)$ is a *quasi-duality module* if the $_S U_R$-duals induce a duality

$$\Delta : \overline{\operatorname{gen}}(U_R) \rightleftarrows \overline{\operatorname{gen}}(_S S) : \Delta$$

between the smallest finitely closed subcategories of Mod-R and S-Mod that contain U_R and $_S S$, respectively.

Just as costar modules are "dual to" $*$-modules, so are their historical predecessors, quasi-duality modules, dual to quasi-progenerators. Here we shall present results on quasi-duality modules that arose from the work of several authors. This section is based on material contained in the papers [39], [40], [41], [42], [59], [61], [70], [83].

We shall soon obtain characterizations of quasi-duality modules that include the fact that they are costar modules. To do so we first need the following notions and results.

A module U is *injective relative to* a module M if $\operatorname{Hom}_R(f, U)$ is an epimorphism whenever $0 \to K \xrightarrow{f} M$ is exact. If U is injective relative to itself, U is called a *quasi-injective* module. According to [1, Propositions 16.13 and 16.10], the class of modules that U is injective relative to is closed under submodules, epimorphic images, and direct sums, and $\prod_A U_\alpha$ is injective relative to M if and only if so is each U_α.

Dual to Lemma 2.4.4 we have

Lemma 4.3.2. *If U is injective relative to each M_i for $i \in I$, and cogenerates all factors of each M_i, then U cogenerates every factor of $\oplus_I M_i$.*

Proof. We first show that if U_R is $M_1 \oplus M_2$-injective and cogenerates all factors of M_1 and of M_2, then U_R cogenerates all factors of $M_1 \oplus M_2$. Suppose $f : M_1 \oplus M_2 \to L$ is an epimorphism and that $0 \neq x \in \mathrm{Rej}_U(L)$. Then $x = f(m_1) + f(m_2)$ where $m_i \in M_i$. If $f(m_1) \in f(M_2)$, then, since $f(M_2)$ is U-torsionless, there is an $\alpha \in \mathrm{Hom}_R(f(M_2), U)$ with $\alpha(x) \neq 0$. But then, since U_R is L-injective, α extends to L, contradicting the choice of x. Hence $f(m_1) \notin f(M_2)$. Thus, since $L/f(M_2)$ is an epimorphic image of M_1 and is hence U-torsionless, there is a $\beta \in \mathrm{Hom}_R(L/f(M_2), U)$ with $\beta(f(m_1) + f(M_2)) \neq 0$, and this induces a map in $\mathrm{Hom}_R(L, U)$, which contradicts the choice of x. It follows that the lemma is true if I is finite. And also, since U is L-injective if $\oplus_{i \in I} M_i \to L$ is an epimorphism, it then follows for arbitrary I. ∎

En route to our characterizations of quasi-duality modules, we determine which quasi-injective modules are costar modules in

Proposition 4.3.3. *A quasi-injective module is a costar module if and only if it cogenerates all its factors.*

Proof. This is immediate from Lemma 4.3.2 and Theorem 4.2.7(e). ∎

As well as being a key to our characterizations of quasi-duality modules, the following lemma will be employed later to characterize (Morita) duality modules.

Lemma 4.3.4. *Suppose that $S = \mathrm{End}(U_R)$ and $_SU$ is a cogenerator. If E_R satisfies $\mathrm{Cogen}(E_R) = \mathrm{Cogen}(U_R)$, then, for any $M \in \mathrm{Mod}\text{-}R$, U_R is M-injective whenever E_R is M-injective. In particular, if $E(U_R)$ is U-torsionless, then U_R is injective.*

Proof. Let Δ denote the $_SU_R$-dual. Fix a monomorphism $U_R \xrightarrow{i} E_R^A$, and for any $\gamma \in \Delta(E_R^A)$ denote $\gamma \circ i \in \mathrm{End}(U_R)$ by s_γ. Note that $\cap \{\mathrm{Ker}\, s_\gamma \,|\, \gamma \in \Delta(E_R^A)\} = 0$. Now suppose $0 \to K_R \xrightarrow{f} M_R$ is exact where E_R is M-injective, and consider the induced exact sequence

$$\Delta(M) \xrightarrow{\Delta(f)} \Delta(K) \longrightarrow C \longrightarrow 0$$

with $C = \mathrm{Coker}\,\Delta(f)$. Suppose $\alpha \in \Delta(K)$. Then, since E_R^A is also M-injective, there is a $\beta : M_R \to E_R^A$ such that

$$
\begin{array}{ccc}
K & \xrightarrow{f} & M \\
\alpha \downarrow & & \beta \downarrow \\
U & \xrightarrow{i} & E^A
\end{array}
$$

commutes. Hence for any $\gamma \in \Delta(E_R^A)$,

$$s_\gamma \alpha = \gamma \circ i \circ \alpha = \gamma \circ \beta \circ f = \Delta f(\gamma \circ \beta).$$

Thus $s_\gamma(\Delta K) \subseteq \operatorname{Im} \Delta f$ and hence $s_\gamma C = 0$ for all $\gamma \in \Delta(E_R^A)$. If $\nu \in \Delta(_SC)$, we now have $s_\gamma(\nu(C)) = \nu(s_\gamma(C)) = 0$ for all $\gamma \in \Delta(E_R^A)$, so $\nu(C) \subseteq \cap\{\operatorname{Ker} s_\gamma | \gamma \in \Delta(E_R^A)\} = 0$. Thus we have shown that $\Delta(_SC) = 0$; therefore, $_SC = 0$ since $_SU$ is a cogenerator. ∎

Now we are ready for

Theorem 4.3.5. *Assume $S = End(U_R)$. The following are equivalent:*

   (a) U_R is a quasi-duality module;
   (b) U_R is a costar module such that every factor of $_SS$ is U-torsionless and every factor of U_R is U-reflexive;
   (c) U_R is quasi-injective and cogenerates all its factors and $_SU$ is an injective cogenerator;
   (d) $Gen(U_R) \subseteq Cogen(U_R)$ and $_SU$ is an injective cogenerator.

Proof. $(a) \Rightarrow (b)$. It suffices to prove that any quasi-duality module is a costar module since the defining duality shows that it satisfies the remaining two conditions. So assume that U_R is a quasi-duality module and let $\mathcal{D}_R$ and $_SC$ be as in Section 4.2. It is immediate that $\Delta : \mathcal{D}_R \to {}_SC$ and that all modules in $_SC$ are U-reflexive. And from this latter fact it follows that $\Delta : {}_SC \to \mathcal{D}_R$. If $M \in \mathcal{D}_R$, then $\Delta(M) \in \overline{\operatorname{gen}}(_SS)$; thus, by hypothesis $\Delta^2(M) \in \overline{\operatorname{gen}}(U_R)$, and since M embeds in $\Delta^2(M)$ it is U-reflexive.

   $(b) \Rightarrow (c)$. Assume that U_R is a costar module satisfying the stated conditions. By Theorem 4.2.7(e) U_R is quasi-injective. Also, $_SU$ is injective because, if $0 \to K \xrightarrow{g} S \to N \to 0$ is exact, then, since $\operatorname{Im} \Delta g$ is a factor of $U_R = \Delta(S)$ and is hence U-reflexive, and since $_SN$ is U-torsionless, we can conclude that Δg is epic by Lemma 4.2.4. Since, also by hypothesis, $_SU$ cogenerates all simple left S-modules, $_SU$ is an injective cogenerator.

   $(c) \Rightarrow (a)$. Assuming (c), U_R is costar by Proposition 4.3.3. If

$$0 \to K_R \to U_R^n \to X_R \to 0$$

is exact, then so is $_SS^n \to {}_S\Delta K \to 0$, and therefore $K \in \mathcal{D}_R$ and U^n are U-reflexive. Thus the exactness properties of Δ_U imply that X is U-reflexive and that $\Delta X \in \overline{\operatorname{gen}}(_SS)$. Now if $M \in \overline{\operatorname{gen}}(U_R)$, we may assume that there is also an exact sequence

$$0 \to M \longrightarrow X \longrightarrow L \to 0$$

with X as above. But then X and L, and hence M, are U-reflexive. Also $\Delta M \in \overline{\text{gen}}(_S S)$, being an epimorphic image of ΔX because U is injective relative to X. If N is a submodule of a finitely generated (U-torsionless, by hypothesis) left S-module Y, then Y and Y/N belong to $_S\mathcal{C}$. Thus the exactness properties of Δ_U yield that N is U-reflexive and that $\Delta N \in \overline{\text{gen}}(U_R)$.

$(d) \Rightarrow (c)$. Let $E_R = \text{Tr}_{U_R}(E(U_R))$. Then E_R is quasi-injective ([1, Exercise 18.17]), so E_R is U_R-injective and clearly $\text{Cogen}(U_R) \subseteq \text{Cogen}(E_R)$. By hypothesis $\text{Cogen}(E_R) \subseteq \text{Cogen}(U_R)$. Hence U_R is quasi-injective by Lemma 4.3.4.

$(c) \Rightarrow (d)$. This follows from Lemma 4.3.2. ∎

The next two results give insight into the structure of the modules involved in a duality between finitely closed subcategories of Mod-R and S-Mod.

Lemma 4.3.6. *Let $\mathcal{A}_R$ and $_S\mathcal{B}$ be full subcategories of Mod-R and S-Mod, respectively, that are closed under finite direct sums and isomorphisms, and suppose that*

$$D : \mathcal{A}_R \rightleftarrows {_S\mathcal{B}} : D$$

is a duality.

(1) If $_S\mathcal{B}$ is closed under epimorphic images and $0 \to K \xrightarrow{f} M$ is exact in $\mathcal{A}_R$, then $DM \xrightarrow{Df} DK \to 0$ is exact in $_S\mathcal{B}$.

(2) If $\mathcal{A}_R$ is closed under submodules and $N \xrightarrow{f} L \to 0$ is exact in $_S\mathcal{B}$, then $0 \to DL \xrightarrow{Df} DN$ is exact in $\mathcal{A}_R$.

Proof. (1) If $DM \xrightarrow{Df} DK \xrightarrow{g} N \to 0$ is exact, then $g \in {_S\mathcal{B}}$ and

$$0 = D(g \circ Df) = DDf \circ Dg.$$

So, since DDf is monic, $Dg = 0$. But D is faithful, so $g = 0$ and Df is epic.

(2) This is dual. ∎

We recall that a module M is *linearly compact* if the inverse limit of any inverse system of epimorphisms $M \xrightarrow{p_\lambda} L_\lambda$ is also an epimorphism. This is equivalent to the condition that any collection of cosets of submodules of M whose finite intersections are all non-empty has a non-empty intersection itself (see [80, 29.7], for example).

Proposition 4.3.7. *If* $\Delta_U : \mathcal{A}_R \rightleftarrows {}_S\mathcal{B} : \Delta_U$ *is a representable duality between finitely closed subcategories of* Mod-R *and* S-Mod, *then*

(1) the modules in $\mathcal{A}_R$ *and* ${}_S\mathcal{B}$ *are linearly compact;*

(2) if $X \in \mathcal{A}_R$ *or* ${}_S\mathcal{B}$, *then the lattices of submodules of* X *and* $\Delta_U X$ *are anti-isomorphic via* $K \mapsto \operatorname{Ker} \Delta_U(i_K)$ *where* $K \xrightarrow{i_K} X$ *is the inclusion map.*

Proof. (1) Suppose that $M \in \mathcal{A}_R$ and

$$0 \to K_\lambda \xrightarrow{i_\lambda} M \xrightarrow{p_\lambda} L_\lambda \to 0$$

is an inverse system of exact sequences. Then by Lemma 4.3.6

$$0 \to \Delta L_\lambda \xrightarrow{\Delta i_\lambda} \Delta M \xrightarrow{\Delta p_\lambda} \Delta K_\lambda \to 0$$

is a direct system of exact sequences in ${}_S\mathcal{B}$. Thus

$$0 \to \varinjlim \Delta L_\lambda \xrightarrow{\varinjlim \Delta i_\lambda} \Delta M \xrightarrow{\varinjlim \Delta p_\lambda} \varinjlim \Delta K_\lambda \to 0$$

is an exact sequence of modules in ${}_S\mathcal{B}$. Now, applying Δ and Lemma 4.3.6 to this sequence, we obtain an exact sequence

$$0 \to \Delta \varinjlim \Delta K_\lambda \xrightarrow{\Delta \varinjlim \Delta i_\lambda} \Delta^2 M \xrightarrow{\Delta \varinjlim \Delta p_\lambda} \Delta \varinjlim \Delta L_\lambda \to 0.$$

But since Δ converts direct limits to inverse limits (see [69, Theorem 2.27] or [80, 29.5]) and M and the L_λ are U-reflexive, it follows that $\varprojlim p_\lambda \cong \Delta \varinjlim \Delta p_\lambda$ is an epimorphism.

(2) We omit this proof, which also employs Lemma 4.3.6. ∎

Now we can characterize quasi-duality modules without reference to their endomorphism rings.

Theorem 4.3.8. *A module* U_R *is a quasi-duality module if and only if* U_R *is quasi-injective, finitely cogenerated, linearly compact, and cogenerates all its factor modules.*

Proof. ($\Rightarrow$). If U_R is a quasi-duality module it is finitely cogenerated by Proposition 4.3.7(2). It is quasi-injective and cogenerates all its factor modules by Theorem 4.3.5. And according to Lemma 4.3.7(1) it is linearly compact.

($\Leftarrow$). Assume that U_R satisfies the conditions. By Proposition 4.3.3, U_R is a costar module inducing a duality $\Delta : \mathcal{D}_R \rightleftarrows {}_S\mathcal{C} : \Delta$ as in Section 4.2. According to Theorem 4.3.5, we need only prove that ${}_S U$ is an injective cogenerator. So let L be a left ideal of S with inclusion map $i : L \to S$, and let $\{L_\alpha \mid \alpha \in A\}$ be the family of finitely generated left ideals contained in L with canonical exact sequences

$$0 \to L_\alpha \xrightarrow{i_\alpha} S \xrightarrow{n_\alpha} S/L_\alpha \to 0.$$

Consider the canonical exact sequences

$$0 \to \Delta(S/L_\alpha) \xrightarrow{\Delta n_\alpha} \Delta S \xrightarrow{\pi} \operatorname{Im} \Delta i_\alpha \to 0$$

and

$$0 \to \operatorname{Im} \Delta i_\alpha \xrightarrow{j} \Delta L_\alpha \xrightarrow{p} \operatorname{Coker} j \to 0.$$

In the latter sequence, $\Delta L_\alpha \in \mathcal{D}_R \subseteq \operatorname{cogen}(U_R)$ and by hypothesis and Lemma 4.3.2 Coker j is U-torsionless; thus, by Proposition 4.2.8, Δj is epic. Now from the commutative diagram

$$
\begin{array}{ccccccccc}
0 \to & L_\alpha & \xrightarrow{i_\alpha} & S & \xrightarrow{n_\alpha} & S/L_\alpha & \to 0 \\
 & \varphi \downarrow & & \delta_S \downarrow & & \delta_{S/L_\alpha} \downarrow \\
0 \to & \Delta \operatorname{Im} \Delta i_\alpha & \xrightarrow{\Delta \pi} & \Delta^2 S & \xrightarrow{\Delta^2 n_\alpha} & \Delta^2 S/L_\alpha
\end{array}
$$

where $\varphi = \Delta j \circ \delta_{L_\alpha}$ is epic and δ_S is an isomorphism, we see that δ_{S/L_α} is monic, that is, S/L_α is U-torsionless and hence belongs to ${}_S\mathcal{C}$. Thus by Corollary 4.2.9 the sequences

$$0 \to \Delta(S/L_\alpha) \xrightarrow{\Delta n_\alpha} \Delta S \xrightarrow{\Delta i_\alpha} \Delta L_\alpha \to 0$$

are exact, and since $\Delta S \cong U_R$ is linearly compact, so are the rows in

$$
\begin{array}{ccc}
\varprojlim \Delta S & \xrightarrow{\varprojlim \Delta i_\alpha} & \varprojlim \Delta L_\alpha \to 0 \\
\cong \downarrow & & \cong \downarrow \\
\Delta S & \xrightarrow{\Delta \varprojlim i_\alpha} & \Delta \varprojlim L_\alpha \to 0.
\end{array}
$$

Thus, since $\varprojlim i_\alpha = i : L \to S$, we see that ${}_S U$ is injective. To show that ${}_S U$ is a cogenerator it suffices to show that for each maximal left ideal L of S, S/L is U-torsionless. But $L = \sum_A L_\alpha$ where each S/L_α is U-torsionless. Thus $\{\mathbf{r}_U(L_\alpha) \mid \alpha \in A\}$ is a family of non-zero submodules of U_R, and since U_R is finitely cogenerated, there is a finite subset $F \subseteq A$ such that

$$\Delta_U(S/L) \cong \mathbf{r}_U(L)) = \cap_A \mathbf{r}_U(L_\alpha) = \cap_F \mathbf{r}_U(L_\alpha) \neq 0. \qquad \blacksquare$$

The following propositions, whose proofs are dual to those of Propositions 2.4.9 and 2.4.10, yield examples of quasi-duality modules.

Proposition 4.3.9. *If U_R is a quasi-duality module with $S = \mathrm{End}(U_R)$ and $B = \mathrm{BiEnd}(U_R)$, then U_B is a quasi-duality module and R is U-dense in B.*

Proposition 4.3.10. *If U_B is a quasi-duality module and R is U-dense in B, then U_R is a quasi-duality module.*

Example 4.3.11. If p is any prime integer, then $\mathbb{Z}_{p^\infty}$ is faithfully balanced and is the minimal injective cogenerator over the linearly compact ring $\mathbb{A}(p)$ of p-adic integers. One can show that any subring of $\mathbb{A}(p)$ is $\mathbb{Z}_{p^\infty}$-dense in $\mathbb{A}(p)$. Thus, in particular the $\mathbb{Z}_{p^\infty}$ is a quasi-duality module over $\mathbb{Z}$.

Examples of different flavors can be found in [39] and [40].

One of the problems that arises naturally when considering a duality represented by a bimodule $_SU_R$ is the characterization of the U-reflexive modules. We conclude this section with a series of results that lead to a characterization of the $_SU_R$-reflexive modules in Mod-R relative to a quasi-duality module U_R with $\mathrm{End}(U_R) = S$. To do so we need a notion more general than linearly compact. (Unfortunately, in this case we know of no characterization of the U-reflexive modules in S-Mod.)

Definition 4.3.12. If M is a module with elements m_λ and submodules K_λ, the family $\{m_\lambda, K_\lambda\}_I$ is *solvable* if there is an $m \in M$ such that $m - m_\lambda \in K_\lambda$ for each $\lambda \in I$. The family $\{m_\lambda, K_\lambda\}_I$ is *finitely solvable* if for each finite subset $F \subseteq I$ there is an $m_F \in M$ such that $m_F - m_\lambda \in K_\lambda$ for each $\lambda \in F$. Note that m is a solution for $\{m_\lambda, K_\lambda\}_I$ if and only if $m \in \cap_I(m_\lambda + K_\lambda)$.

The connection between solvability and inverse limits is given in

Lemma 4.3.13. *Let $\{(L_\lambda), (f_{\lambda\mu})\}_{(I,\leq)}$ be an inverse system of modules with inverse system of epimorphisms $p_\lambda : M \to L_\lambda$ (so that $f_{\lambda\mu} \circ p_\mu = p_\lambda$ whenever $\lambda \leq \mu$ in I). If $K_\lambda = \mathrm{Ker}\, p_\lambda$ $(\lambda \in I)$, then $\varprojlim p_\lambda : M \to \varprojlim L_\lambda$ is an epimorphism if and only if every finitely solvable family $\{m_\lambda, K_\lambda\}_I$ in M is solvable.*

Proof. We may consider $\varprojlim L_\lambda = \{(\ell_\lambda)_I \mid f_{\lambda\mu}(\ell_\mu) = \ell_\lambda\}$ whenever $\lambda \leq \mu$ in I and $\varprojlim p_\lambda(m) = (p_\lambda(m))_I$ ([69, Theorem 2.22 and page 55]).

($\Rightarrow$). Consider $(p_\lambda(m_\lambda))_I \in \prod_I L_\lambda$. If $\lambda \leq \mu$ in I, then, by assumption there is an $m_0 \in (m_\lambda + K_\lambda) \cap (m_\mu + K_\mu)$. But then

$$f_{\lambda\mu}(p_\mu(m_\mu)) = f_{\lambda\mu}(p_\mu(m_0)) = p_\lambda(m_0) = p_\lambda(m_\lambda).$$

Thus $(p_\lambda(m_\lambda))_I \in \varprojlim L_\lambda$, so since we are assuming that $\varprojlim p_\lambda : M \to \varprojlim L_\lambda$ is an epimorphism, there is an $m \in M$ such that $p_\lambda(m) = p_\lambda(m_\lambda)$ for all $\lambda \in I$, that is, $m \in \cap_I (m_\lambda + K_\lambda)$.

($\Leftarrow$). Suppose that $(\ell_\lambda)_I = (p_\lambda(m_\lambda))_I \in \varprojlim L_\lambda$. If F is a finite subset of I, let $\lambda \leq k \in F$ for all $\lambda \in F$. Then

$$p_\lambda(m_\lambda) = \ell_\lambda = f_{\lambda k}(\ell_k) = f_{\lambda k} \circ p_k(m_k) = p_\lambda(m_k),$$

so $m_k \in \cap_F (m_\lambda + K_\lambda)$. Thus by hypothesis there is an $m \in M$ such that $p_\lambda(m) = \ell_\lambda$ for all $\lambda \in I$. ∎

Taken together, the notions in the following definition suffice to ensure reflexivity.

Definition 4.3.14. Let $_S U_R$ be a bimodule. If $M \in \text{Mod-}R$, then

(1) M is *U-dense* (in $\Delta^2 M$) if for each $h \in \Delta^2 M$ and each finite set $F \subseteq \Delta M$ there is an $m \in M$ such that $h(f) = f(m)$ for all $f \in F$;
(2) M is *U-linearly compact* if M is *U*-torsionless, and $\varprojlim p_\lambda : M \to \varprojlim L_\lambda$ is an epimorphism for every inverse system of epimorphisms $p_\lambda : M \to L_\lambda$ with each L_λ *U*-torsionless.

Note that if $S = \text{End}(U_R)$, then $\text{BiEnd}(U_R) \cong \Delta^2(R_R)$, via the canonical isomorphism $U_R \cong \text{Hom}_R(R, U)$, and so R_R is *U*-dense if and only if R is *U*-dense in $B = \text{BiEnd}(U_R)$.

It follows from Lemma 4.3.13 that, as observed in [41], M is *U*-linearly compact if and only if every finitely solvable family $\{m_\lambda, K_\lambda\}_I$ with each $m_\lambda \in M$ and $M/K_\lambda \in \text{Cogen}(U_R)$ is solvable.

If $\{m_\lambda, K_\lambda\}_I$ is a finitely solvable family in a module M with $m_F - m_\lambda \in K_\lambda$ for each finite subset $F \subseteq I$, then, letting $K_F = \cap_{\lambda \in F} K_\lambda$, one checks that, if $\mathcal{I}$ denotes the set of finite subsets of I, then $\{m_F, K_F\}_\mathcal{I}$ is finitely solvable. Indeed, if $F_1, \ldots, F_n$ are finite subsets of I, and $m_{\cup_{j=1}^n F_j} - m_\lambda \in K_\lambda$, for $\lambda \in \cup_{j=1}^n F_j$ then $m_{\cup_{j=1}^n F_j} - m_{F_\ell} = m_{\cup_{j=1}^n F_j} - m_\lambda + m_\lambda - m_{F_\ell} \in K_{F_\ell}$ for $\ell = 1, \ldots, n$.

Lemma 4.3.15. *If $_S U_R$ is a bimodule and M_R is U-dense and U-linearly compact, then M is U-reflexive.*

Proof. Let $h \in \Delta^2 M$. Since M is *U*-dense there is, in particular, for each $f \in \Delta M$ an $m_f \in M$ with $h(f) = f(m_f)$. Consider $\{m_f + \text{Ker } f \mid f \in \Delta M\}$. If F is a finite subset of ΔM, then there is an m_F such that $h(f) = f(m_F)$

for all $f \in F$ and so

$$f(m_F) = h(f) = f(m_f).$$

Thus $\{m_f, \operatorname{Ker} f\}_{\Delta M}$ is finitely solvable. But then for each finite subset $F \subseteq \Delta M$ we let $m_F - m_f \in \operatorname{Ker} f$ for $f \in F$ and $K_F = \cap_F \operatorname{Ker} f$ so that the system $\{m_F, K_F\}_{\mathcal{I}}$ is finitely solvable (where $\mathcal{I}$ denotes the set of finite subsets of ΔM). Now $M/K_F \in \operatorname{Cogen}(U_R)$ and the canonical maps $p_F : M \to M/K_F$ form an inverse system of epimorphisms. Thus, since M is U-linearly compact, there is by Lemma 4.3.13 an $m \in M$ such that $m - m_F \in K_F$ for all $F \subseteq \Delta M$, and so if $f \in F$,

$$\delta_M(m)(f) = f(m) = f(m_F) = f(m_f) = h(f);$$

therefore, δ_M is an isomorphism. ∎

Lemma 4.3.16. *If* $\operatorname{gen}(U_R) \subseteq \operatorname{Cogen}(U_R)$, *then every right R-module is U-dense.*

Proof. This is proved in [80, 47.6(4)] and is a part of the proof of [81, Theorem 4.1]. ∎

Also connecting U-reflexivity and U-linearly compactness, we have

Lemma 4.3.17. *Let $_SU_R$ be a bimodule. If M_R is a U-reflexive module such that $_SU$ is ΔM-injective, then M is U-linearly compact.*

Proof. Let $M \xrightarrow{p_\lambda} L_\lambda \to 0$ be an inverse system of epimorphisms with the $L_\lambda \in \operatorname{Cogen}(U_R)$, and apply Δ to obtain a direct system of monomorphisms $0 \to \Delta L_\lambda \xrightarrow{\Delta p_\lambda} \Delta M$. Then

$$0 \to \varinjlim \Delta L_\lambda \xrightarrow{\varinjlim \Delta p_\lambda} \Delta M$$

is exact, and so since $_SU$ is ΔM-injective, we obtain a commutative diagram with exact top row

$$
\begin{array}{ccc}
\Delta^2 M & \xrightarrow{\Delta(\varinjlim \Delta p_\lambda)} & \Delta(\varinjlim \Delta L_\lambda) \quad \to 0 \\
\| & & \cong \uparrow \\
\Delta^2 M & \xrightarrow{\varprojlim(\Delta^2 p_\lambda)} & \varprojlim(\Delta^2 L_\lambda) \\
\cong \uparrow & & \varprojlim(\delta_{L_\lambda}) \uparrow \\
M & \xrightarrow{\varprojlim p_\lambda} & \varprojlim L_\lambda
\end{array}
$$

from which, since $\varprojlim(\delta_{L_\lambda})$ is a monomorphism, it follows that $\varprojlim p_\lambda$ is an epimorphism. ■

Finally, we have the promised characterization.

Proposition 4.3.18. *If U_R is a quasi-duality module, then M_R is U-reflexive if and only if M is U-linearly compact.*

Proof. Since U is injective over $S = \operatorname{End}(U_R)$, the condition is necessary by Lemma 4.3.17. It is sufficient by Lemmas 4.3.16 and 4.3.15. ■

4.4. Morita Duality

The various forms of duality theories for modules sprang from what has come to be known as Morita duality, which originated in the papers of G. Azumaya [6] and K. Morita [65]. Most of the results in this brief section, and several further details about Morita duality, can be found in [1] and [81].

Definition 4.4.1. A module U_R with $S = \operatorname{End}(U_R)$ is a *(Morita) duality module* if U_R is faithful and balanced (so that $R \cong \operatorname{End}(_S U)$, canonically), and the $_S U_R$-reflexive modules are closed under submodules and epimorphic images. In this case the faithfully balanced bimodule $_S U_R$ is said to induce a *Morita duality*.

Note that a bimodule $_S U_R$ is faithfully balanced if and only if R_R and $_S S$ are U-reflexive.

Theorem 4.4.2. *The following statements about a bimodule $_S U_R$ are equivalent.*

  (a) $_S U_R$ induces a Morita duality;
  (b) Every epimorphic image of R_R, $_S S$, U_R, and $_S U$ is U-reflexive;
  (c) $_S U_R$ is faithfully balanced and U_R and $_S U$ are injective cogenerators;
  (d) $_S U_R$ is faithfully balanced and U_R and $_S U$ are cogenerators;

Proof. $(a) \Rightarrow (b)$. This is clear.
   $(b) \Rightarrow (c)$. Since every cyclic module and every epimorphic image of U is U-reflexive, and hence U-torsionless, we can apply Lemma 4.2.4 to an exact sequence $0 \to I \xrightarrow{f} R_R \longrightarrow R/I \to 0$ to see that Δf is epic. Thus U_R is injective and a cogenerator.

$(c) \Rightarrow (a)$. This follows easily from the implied exactness of Δ^2.
$(c) \Rightarrow (d)$ is obvious and $(d) \Rightarrow (c)$ follows from Lemma 4.3.4. ∎

From Theorem 4.3.5 and Theorem 4.4.2 we obtain

Corollary 4.4.3. *A module U_R is a duality module if and only if it is a quasi-duality module that is faithfully balanced and a cogenerator.*

If $_SU_R$ is a duality module, then both U_R and $_SU$ are quasi-duality modules. Thus from Proposition 4.3.18 we have the following result of B.J. Müller [66].

Corollary 4.4.4. *If $_SU_R$ induces a Morita duality, then the U-reflexive modules are precisely the linearly compact modules in* Mod-R *and* S-Mod *.*

Corollary 4.4.5. *A module U_R is a duality module if and only if U_R is a balanced linearly compact finitely cogenerated injective cogenerator.*

Proof. By Theorems 4.3.8 and 4.3.5, if $S = \text{End}(U_R)$, then $_SU$ is an injective cogenerator ∎

Corollary 4.4.6. *A module U_R is a duality module if and only if R_R and U_R are linearly compact and U_R is a finitely cogenerated injective cogenerator.*

Proof. If U_R is a duality module, then by Corollary 4.4.5 we need only note that R_R is linearly compact, and this follows from Theorem 4.4.2 and Corollary 4.4.4. Conversely, U_R is a quasi-duality module by Theorem 4.3.8. Thus, since R_R is linearly compact, R_R is U-reflexive by Proposition 4.3.18, so U_R is faithfully balanced and Corollary 4.4.3 applies. ∎

5

Cotilting

We consider dual notions of tilting modules and the Tilting Theorem. We begin by examining a type of theorem that is dual to the Tilting Theorem in a manner similar to the way Morita duality is related to Morita equivalence.

We continue our practice of using the term *subcategory* of Mod-R or S-Mod to indicate a full subcategory that is closed under isomorphisms. And we say that an *abelian subcategory* of Mod-R or S-Mod is a full subcategory that is closed under finite direct sums and contains the kernels and cokernels of all of its homomorphisms.

In addition to the contravariant Hom functors

$$\Delta_{U_R} : \text{Mod-}R \rightleftarrows S\text{-Mod} : \Delta_{sU}$$

(both of which we usually denote by Δ) derived from a bimodule $_sU_R$, henceforth, we shall let $\Gamma_{U_R} = \text{Ext}^1_R(\_, U)$ and $\Gamma_{sU} = \text{Ext}^1_S(\_, U)$ to obtain another pair of contravariant functors

$$\Gamma_{U_R} : \text{Mod-}R \rightleftarrows S\text{-Mod} : \Gamma_{sU},$$

both of which will usually be denoted by Γ.

5.1. Cotilting Theorem

As we shall see, there are several versions of "cotilting" modules that lead to a "cotilting theorem" as defined here.

Definition 5.1.1. Let $\mathcal{A}_R$ and $_s\mathcal{A}$ be abelian subcategories of Mod-R and of S-Mod, respectively, such that $R_R \in \mathcal{A}_R$ and $_sS \in {}_s\mathcal{A}$. Let $_sU_R$ be a bimodule, and let

$$\mathcal{T}_R = \text{Ker}\,\Delta \cap \mathcal{A}_R, \quad \mathcal{F}_R = \text{Ker}\,\Gamma \cap \mathcal{A}_R,$$
$$_s\mathcal{T} = \text{Ker}\,\Delta \cap {}_s\mathcal{A}, \quad _s\mathcal{F} = \text{Ker}\,\Gamma \cap {}_s\mathcal{A}.$$

86

Then $_sU_R$ induces a *cotilting theorem* between $\mathcal{A}_R$ and $_s\mathcal{A}$ if the following four conditions are satisfied:

(1) $(\mathcal{T}_R, \mathcal{F}_R)$ and $(_s\mathcal{T}, _s\mathcal{F})$ are torsion theories in $\mathcal{A}_R$ and $_s\mathcal{A}$, respectively;

(2) $\Delta : \mathcal{A}_R \to _s\mathcal{F}$, $\Gamma : \mathcal{A}_R \to _s\mathcal{T}$, $\Delta : _s\mathcal{A} \to \mathcal{F}_R$, $\Gamma : _s\mathcal{A} \to \mathcal{T}_R$;

(3) There are natural transformations $\gamma : \Gamma^2 \to 1_{\mathcal{A}_R}$ and $\gamma : \Gamma^2 \to 1_{_s\mathcal{A}}$ that, together with the evaluation maps $\delta : 1_{\mathcal{A}_R} \to \Delta^2$ and $\delta : 1_{_s\mathcal{A}} \to \Delta^2$, yield exact sequences

$$0 \to \Gamma^2 M \xrightarrow{\gamma_M} M \xrightarrow{\delta_M} \Delta^2 M \to 0 \quad \text{and}$$
$$0 \to \Gamma^2 N \xrightarrow{\gamma_N} N \xrightarrow{\delta_N} \Delta^2 N \to 0$$

for each $M \in \mathcal{A}_R$ and each $N \in _s\mathcal{A}$;

(4) The restrictions

$$\Delta : \mathcal{F}_R \rightleftarrows _s\mathcal{F} : \Delta \quad \text{and} \quad \Gamma : \mathcal{T}_R \rightleftarrows _s\mathcal{T} : \Gamma$$

define category equivalences.

Suppose that $_sU_R$ induces a cotilting theorem between $\mathcal{A}_R$ and $_s\mathcal{A}$. Then, since the abelian subcategories $\mathcal{A}_R$ and $_s\mathcal{A}$ contain R_R and $_sS$, respectively, they also contain all finitely presented modules. Also $_sU \cong \Delta(R_R) \in _s\mathcal{F}$ and $U_R \cong \Delta(_sS) \in \mathcal{F}_R$ so that $\text{Ext}_S^1(U, U) = 0$ and $\text{Ext}_R^1(U, U) = 0$, and $_sU_R$ is faithfully balanced since $R_R \in \mathcal{F}_R$ and $_sS \in _s\mathcal{F}$ are U-reflexive. Also dual to Theorem 3.2.1 (c), $\text{Ker}\,\Delta \cap \text{Ker}\,\Gamma \cap \mathcal{A}_R = 0$ and $\text{Ker}\,\Delta \cap \text{Ker}\,\Gamma \cap _s\mathcal{A} = 0$.

Clearly, condition (4) of Definition 5.1.1 follows from the other conditions. In this section we employ a series of lemmas to show that, given some further restrictions, condition (3) follows from (1), (2), and the first part of (4).

For the remainder of this section we assume that $\mathcal{A}_R$ and $_s\mathcal{A}$ are abelian subcategories of Mod-R and S-Mod. Moreover, we shall use $\mathcal{A}$ to denote both $\mathcal{A}_R$ and $_s\mathcal{A}$, etc.

Lemma 5.1.2. *Suppose $_sU_R$ satisfies conditions (1) and the first half of (4) of Definition 5.1.1. If $X \in \mathcal{A}$, then δ_X is epic and X has torsion subobject (torsion submodule) $\tau(X) = \text{Ker}\,\delta_X$.*

Proof. Let $X \in \mathcal{A}$ and let $T \in \mathcal{T}$ be a submodule of X such that $X/T \in \mathcal{F}$. If $p : X \to X/T$ is the natural epimorphism, then $\Delta(p)$ is an isomorphism by hypothesis, so $\Delta^2(p)$ is an isomorphism. Also $\delta_{X/T}$ is an isomorphism by hypothesis (see Proposition 4.1.2). Since p is an epimorphism, the conclusions follow from the commutative diagram

$$
\begin{array}{ccc}
X & \xrightarrow{\ p\ } & X/T & \to 0 \\
\delta_X \downarrow & & \delta_{X/T} \downarrow & \qquad \blacksquare \\
\Delta^2 X & \xrightarrow{\Delta^2(p)} & \Delta^2(X/T). &
\end{array}
$$

Lemma 5.1.3. *Suppose $_S U_R$ satisfies conditions (1), (2), and the first half of (4) of Definition 5.1.1. If $Y \in \mathcal{T} \cap \mathrm{gen}(\mathcal{F})$, then there is an X in $\mathcal{F}$ and a commutative diagram*

$$
\begin{array}{ccccccc}
0 \to & L & \xrightarrow{\ i\ } & X & \xrightarrow{\ p\ } & Y & \to 0 \\
 & \delta_L \downarrow & & \delta_X \downarrow & & \nu_Y \downarrow & \\
0 \to & \Delta^2 L & \xrightarrow{\Delta^2 i} & \Delta^2 X & \xrightarrow{\ \partial\ } & \Gamma^2 Y & \to 0
\end{array}
$$

with exact rows in which all vertical maps are isomorphisms.

Proof. Let $X \in \mathcal{F}$ and suppose

$$
0 \to L \xrightarrow{\ i\ } X \xrightarrow{\ p\ } Y \to 0
$$

is exact with $Y \in \mathcal{T}$. Then by definition of $\mathcal{T}$ and $\mathcal{F}$, we have an exact sequence

$$
0 \to \Delta X \xrightarrow{\Delta i} \Delta L \xrightarrow{\partial_1} \Gamma Y \to 0.
$$

But by (2), $\Delta \Gamma Y = 0 = \Gamma \Delta L$, so this sequence yields the desired bottom row. Commutativity of the left square and exactness of the rows allow us to define $\nu_Y = \partial \circ \delta_X \circ p^{-1}$. By hypothesis and Lemma 5.1.2, both δ_X and δ_L are isomorphisms. Thus, so is ν_Y. $\blacksquare$

Lemma 5.1.4. *Suppose $\mathcal{F}$ is closed under extensions of modules. If $0 \to L \xrightarrow{\ f\ } X \xrightarrow{\ f\ } Y \to 0$ is exact with $L \in \mathcal{F}$ and $Y \in \mathrm{gen}(\mathcal{F})$, then $X \in \mathrm{gen}(\mathcal{F})$.*

Proof. There is an $H \in \mathcal{F}$, an epimorphism g and a pullback of f and g that yield a commutative diagram

$$
\begin{array}{ccccccc}
0 \to & L & \to & Q & \to & H & \to 0 \\
 & \| & & \downarrow & & g \downarrow & \\
0 \to & L & \to & X & \xrightarrow{\ f\ } & Y & \to 0 \\
 & & & \downarrow & & \downarrow & \\
 & & & 0 & & 0 &
\end{array}
$$

with exact rows and columns. Thus, since $\mathcal{F}$ is closed under extensions, $X \in \text{gen}(\mathcal{F})$. ∎

Lemma 5.1.5. *Suppose $\mathcal{F}$ is closed under extensions of modules and $_S U_R$ satisfies conditions (1), (2), and the first half of condition (4) of Definition 5.1.1. The collection of mappings $\nu = (\nu_Y)$ is a natural isomorphism on the full category $\mathcal{T} \cap \text{gen}(\mathcal{F})$.*

Proof. Given a map $f : Y' \to Y$ between modules in $\tau \cap \text{gen}(\mathcal{F})$, there is a pull-back diagram

$$\begin{array}{ccccccccc} 0 & \to & L & \overset{j}{\to} & P & \overset{n_1}{\to} & Y' & \to & 0 \\ & & \| & & f_1 \downarrow & & f \downarrow & & \\ 0 & \to & L & \overset{i}{\to} & X & \overset{n}{\to} & Y & \to & 0 \end{array}$$

with $X \in \mathcal{F}$. Now by Lemma 5.1.4, there is an epimorphism $X' \overset{g}{\to} P \to 0$ with $X' \in \mathcal{F}$. Then, letting $L' = \text{Ker}\, n_1 \circ g \leq X'$, we obtain a commutative diagram with exact rows

$$\begin{array}{ccccccccc} 0 & \to & L' & \overset{i'}{\to} & X' & \overset{n'}{\to} & Y' & \to & 0 \\ & & \downarrow & & f' \downarrow & & f \downarrow & & \\ 0 & \to & L & \overset{i}{\to} & X & \overset{n}{\to} & Y & \to & 0 \end{array}$$

in which

$$n' = n_1 \circ g \quad \text{and} \quad f' = f_1 \circ g.$$

Thus, since $\Delta Y = 0 = \Gamma X$ and $\Delta Y' = 0 = \Gamma X'$, this yields a commutative diagram with exact rows

$$\begin{array}{ccccccc} 0 \to & \Delta X & \overset{\Delta i}{\longrightarrow} & \Delta L & \longrightarrow & \Gamma Y & \to 0 \\ & \Delta f' \downarrow & & \downarrow & & \Gamma f \downarrow & \\ 0 \to & \Delta X' & \overset{\Delta i}{\longrightarrow} & \Delta L' & \longrightarrow & \Gamma Y' & \to 0 \end{array},$$

which, in particular, yields a commutative diagram

$$\begin{array}{ccc} \Delta^2 X & \overset{\partial}{\to} & \Gamma^2 Y \\ \Delta^2 f' \uparrow & & \Gamma^2 f \uparrow \\ \Delta^2 X' & \overset{\partial'}{\to} & \Gamma^2 Y' \end{array}$$

(see [69, Theorem 6.4, p. 173]). Now, defining $\nu_{Y'}$ via the exact sequence

$$0 \to L' \longrightarrow X' \longrightarrow Y' \to 0,$$

we obtain a cube

$$
\begin{array}{ccccc}
 & X & \xrightarrow{\;n\;} & Y & \longrightarrow 0 \\
 {}^{f'}\nearrow \quad | & & {}^{f}\nearrow \quad \nu_Y| & & \\
 X' & \xrightarrow{\;n'\;} & Y' & \longrightarrow 0 & \\
 \downarrow & & \downarrow & & \\
 \downarrow \quad \Delta^2 X & \xrightarrow{\;-\nu_{Y'}\;}\downarrow & \to & \Gamma^2 Y & \\
 & & {}_{\Gamma^2 f} & & \\
 \Delta^2 X' & \longrightarrow & \Gamma^2 Y' & &
\end{array}
$$

in which all squares, except possibly the right one, commute. Then, since n is epic, so does the desired right square commute.

Finally, since the sequence defining $\nu_{Y'}$ was derived from the sequence defining ν_Y, we need to show that, if ν_Y is derived from $0 \to L \longrightarrow X \xrightarrow{f} Y \to 0$ and ν'_Y is derived from $0 \to L' \longrightarrow X' \xrightarrow{f'} Y \to 0$, then $\nu_Y = \nu'_Y$. So, following [58, Proposition 9], we employ the commutative pull-back diagram with exact rows, columns, and diagonal

$$
\begin{array}{ccccc}
 & 0 & 0 & 0 & \\
 & \uparrow & \uparrow & \nearrow & \\
 0 \to L \to & X & \xrightarrow{f} & Y & \to 0 \\
 \| & {}^{g'}\uparrow & {}^{h}\nearrow \; {}^{f'}\uparrow & & \\
 0 \to L \to & P & \xrightarrow{g} & X' & \to 0 \\
 & \nearrow \uparrow & & \uparrow & \\
 K & L' & = & L' & \\
 \nearrow & \uparrow & & \uparrow & \\
 0 & 0 & & 0 &
\end{array}
$$

with $P \in \mathcal{F}$ since $L, X' \in \mathcal{F}$. Now let ν_Y^0 be derived from $0 \to K \longrightarrow P \xrightarrow{h} Y \to 0$ so the commutative square

$$
\begin{array}{ccc}
 X & \xrightarrow{f} & Y \\
 {}^{g'}\uparrow & & {}^{1_Y}\uparrow \\
 P & \xrightarrow{h} & Y
\end{array}
$$

yields (see the cube) $\nu_Y = \nu_Y \circ 1_Y = 1_{\Gamma^2 Y} \circ \nu_Y^0 = \nu_Y^0$. Similarly, $\nu_Y^0 = \nu'_Y$. ∎

Finally we can show how condition (3) follows from the other conditions and ones that we shall see are satisfied by various versions of cotilting modules.

Proposition 5.1.6. *Suppose $_sU_R$ satisfies conditions (1), (2), and the first half of condition (4) of Definition 5.1.1. If, in addition, both $\mathcal{F}_R$ and $_s\mathcal{F}$ are closed under extensions in* Mod-R *and* S-Mod, $\mathcal{T}_R \subseteq \mathrm{gen}(\mathcal{F}_R)$ *and* $_s\mathcal{T} \subseteq \mathrm{gen}(_s\mathcal{F})$, *and* $_sU_R$ *satisfies* $\mathrm{Ext}^2(\_, U) = 0$ *on* $\mathcal{F}_R$ *and* $_s\mathcal{F}$, *then condition (3) is also satisfied, that is, $_sU_R$ induces a cotilting theorem between* $\mathcal{A}_R$ *and* $_s\mathcal{A}$.

Proof. According to Lemma 5.1.5, the additional conditions imply that all of condition (4) is satisfied. Let $X \in \mathcal{A}$. By Lemma 5.1.2, if $T = \mathrm{Ker}\,\delta_X$, then T is a torsion module in $\mathcal{A}$. Also, if $i_T : T \to X$ is the inclusion map, then, since $\Gamma\Delta^2 X = 0$, we see from

$$0 \to T \xrightarrow{i_T} X \xrightarrow{\delta_X} \Delta^2 X \to 0$$

that

$$0 \to \Gamma X \xrightarrow{\Gamma i_T} \Gamma T \to 0$$

is exact, and so $\Gamma^2 i_T$ is an isomorphism. Now employing (4) and noting that if $f : X \to X'$, then $f : \mathrm{Ker}\,\delta_X \to \mathrm{Ker}\,\delta_{X'}$, we see that $\gamma_X = i_T \circ \nu_T^{-1} \circ \Gamma^2(i_T)^{-1}$ is the desired natural transformation. Indeed, if $f : X \to X,'$ then all the squares commute in

$$
\begin{array}{ccccccc}
X & \xleftarrow{i} & T & \xrightarrow{\nu_T} & \Gamma^2 T & \xrightarrow{\Gamma^2 i_T} & \Gamma^2 X \\
f \downarrow & & \overline{f} \downarrow & & \Gamma^2 \overline{f} \downarrow & & \Gamma^2 f \downarrow \\
X' & \xleftarrow{i'} & T' & \xrightarrow{\nu_{T'}} & \Gamma^2 T' & \xrightarrow{\Gamma^2 i_{T'}} & \Gamma^2 X'. \quad \blacksquare
\end{array}
$$

5.2. Cotilting Modules

In this section we discuss the dual version of generalized tilting modules, which we shall simply refer to as cotilting modules. In the next section we shall establish a cotilting theorem for so-called cotilting bimodules and employ a principal result of this section to characterize their reflexive modules.

We begin with a result dual to one that was valuable in our study of generalized tilting modules. For any R module U we denote the kernel of $\mathrm{Ext}^1_R(\_, U)$ by $^\perp U$.

Proposition 5.2.1. $^\perp U$ *is closed under submodules if and only if* inj . dim .U ≤ 1.

Proof. This is dual to the proof of Proposition 1.1.1. $\blacksquare$

Now, as promised, we have the definition of a cotilting module. It first appeared in the work of R. Colpi, G. D'Este, and A. Tonolo [28], where they proved the characterization in Proposition 5.2.6 to follow.

Definition 5.2.2. A module U_R is a *cotilting module* if $\mathrm{Cogen}(U_R) = {}^{\perp}U$.

For a bimodule ${}_SU_R$, we shall, when convenient, continue to denote the contravariant functors $\mathrm{Hom}_R(\_, U)$ and $\mathrm{Hom}_S(U, \_)$ by Δ and $\mathrm{Ext}^1_R(\_, U)$ and $\mathrm{Ext}^1_S(\_, U)$ by Γ, adding subscripts when necessary.

If U_R is a cotilting module with $S = \mathrm{End}(U_R)$, and $0 \to M \xrightarrow{f} X \longrightarrow L \to 0$ is exact with X_R a ${}_SU_R$-reflexive module (for example $X = U^n$), then we have an exact sequence $\Delta X \xrightarrow{\Delta f} \Delta M \to \Gamma L \to \Gamma X = 0$, from which we see that Δf is epic if and only if $L \in {}^{\perp}U = \mathrm{Cogen}(U_R)$. Thus, employing Theorem 4.2.7(e), we obtain

Proposition 5.2.3. *Every cotilting module is a costar module.*

Dual to Proposition 3.1.3 we have

Proposition 5.2.4. *If U_R is a cotilting module, then*

$$\mathrm{Cogen}(U_R) = \mathrm{Copres}(U_R).$$

Proof. Let $M \in \mathrm{Cogen}(U_R)$ (so that $M = M/\mathrm{Rej}_U(M)$) and apply Lemma 4.2.1. ∎

Now we can verify duals to Theorem 3.1.4 and Proposition 3.1.5 to obtain the following two characterizations of cotilting modules. The first of these was obtained by L. Angeleri Hügel, A. Tonolo, and J. Trlifaj in [3].

Theorem 5.2.5. *A module U_R is a cotilting module if and only if*

(i) $\mathrm{inj}.\dim.U \leq 1$,
(ii) $\mathrm{Ext}^1_R(U^A, U) = 0$ *for all sets A,*
(iii) *an injective cogenerator C_R admits an exact sequence $0 \to U_1 \longrightarrow U_0 \longrightarrow C \to 0$ with $U_0, U_1 \in \mathrm{Prod}(U_R)$.*

Proof. Assume that U_R is a cotilting module.

(i) follows from Proposition 5.2.1.
(ii) is obvious from the definition.

(iii) Since $R_R \in {}^{\perp}U = \mathrm{Cogen}(U_R)$, it follows that every injective right R-module is an epimorphic image of a direct product of copies of U. Thus there is an exact sequence

$$0 \to K \xrightarrow{f} U^A \longrightarrow C \to 0,$$

and by Proposition 5.2.4 another exact sequence

$$0 \to K \xrightarrow{g} U^B \longrightarrow L \to 0$$

with $L \in \mathrm{Cogen}(U_R)$. An argument dual to that of Theorem 3.1.4 using the pushout diagram of f and g completes the proof of this implication.

Conversely, assume (i), (ii), and (iii). If $M \in \mathrm{Cogen}(U_R)$, then an exact sequence $0 \to M \longrightarrow U^A \longrightarrow L \to 0$ yields an exact sequence

$$0 = \mathrm{Ext}^1_R(U^A, U) \to \mathrm{Ext}^1_R(M, U) \to \mathrm{Ext}^2_R(L, U) = 0.$$

On the other hand, assume that $M \in {}^{\perp}U$ and let $p : M \to M/Rej_{U_0}(M)$ be the canonical epimorphism to obtain, from (iii), a commutative diagram

$$
\begin{array}{ccc}
\mathrm{Hom}_R(M, U_0) & \xrightarrow{\alpha} & \mathrm{Hom}_R(M, C) \\
\mathrm{Hom}_R(p, U_0) \uparrow & & \mathrm{Hom}_R(p, C) \uparrow \\
\mathrm{Hom}_R(M/Rej_{U_0}(M), U_0) & \longrightarrow & \mathrm{Hom}_R(M/Rej_{U_0}(M), C)
\end{array}
$$

in which $\mathrm{Hom}_R(p, U_0)$ is an isomorphism by definition of $Rej_{U_0}(M)$, and α is an epimorphism since $U_1 \in \mathrm{Prod}(U_R)$. But now it follows that $\mathrm{Hom}_R(p, C)$ is an isomorphism, and since C is a cogenerator, so is p. Thus $M \in \mathrm{Cogen}(U_0) \subseteq \mathrm{Cogen}(U)$. ∎

Proposition 5.2.6. *A module U_R is a cotilting module if and only if*

   (i) inj.dim.$U \le 1$;
   (ii) $\mathrm{Ext}^1_R(U^A, U) = 0$ *for all sets A;*
   (iii) $\mathrm{Ker}\,\mathrm{Hom}_R(\_, U) \cap {}^{\perp}U = 0$.

Proof. If U is cotilting, then (i) and (ii) follow from Theorem 5.2.5, and if $\mathrm{Hom}_R(M, U) = 0 = \mathrm{Ext}^1_R(M, U)$, then $M \in {}^{\perp}U = \mathrm{Cogen}(U)$, so $M = 0$ and (iii) is verified.

Conversely, ${}^{\perp}U$ is closed under submodules by (i) and Proposition 5.2.1, so by (ii) $\mathrm{Cogen}(U_R) \subseteq {}^{\perp}U$. If $M \in {}^{\perp}U$, then from

$$0 \to \mathrm{Rej}_U(M) \longrightarrow M \longrightarrow M/\mathrm{Rej}_U(M) \to 0$$

we obtain the exact sequences

$$\Delta(M/\operatorname{Rej}_U(M)) \xrightarrow{\cong} \Delta M \to \Delta(\operatorname{Rej}_U(M)) \to \Gamma(M/\operatorname{Rej}_U(M)) = 0$$

and

$$\Gamma M \to \Gamma(\operatorname{Rej}_U(M)) \to \operatorname{Ext}^2_R(M/\operatorname{Rej}_U(M), U) = 0,$$

from which we see that $\Delta(\operatorname{Rej}_U(M)) = 0 = \Gamma(\operatorname{Rej}_U(M))$. Thus by condition (iii), $^\perp U \subseteq \operatorname{Cogen}(U_R)$. ■

As one might hope, if R is an artin algebra, finitely generated cotilting modules are just the artin algebra duals of tilting modules.

Proposition 5.2.7. *Let R be an artin algebra with D the artin algebra dual. Let U_R be finitely generated. Then U_R is a cotilting module if and only if $_RV = DU$ is a tilting module.*

Proof. ($\Rightarrow$) If $M \in \operatorname{gen}(_RV)$, then $DM \in \operatorname{cogen}(U_R) \subseteq {^\perp U}$. But then $\operatorname{Ext}^1_R(V, M) \cong \operatorname{Ext}^1_R(DM, U) = 0$, so $M \in V^\perp \cap R\text{-mod}$. On the other hand, if $M \in V^\perp \cap R\text{-mod}$, then $DM \in {^\perp U} = \operatorname{Cogen}(U_R)$, and so, being finitely cogenerated, $DM \in \operatorname{cogen}(U_R)$, that is, $M \in \operatorname{gen}(_RV)$. Thus $_RV = DU$ is a tilting module by Proposition 3.2.3.

($\Leftarrow$) If $_RV = DU$ is a tilting module, then $U_R \cong DV$ clearly satisfies conditions (i) and (iii) of Theorem 5.2.5, and by Corollary 1.3.3 $U^A \in \operatorname{Add}(U_R)$, so, since $\operatorname{Ext}^1_R(U^{(B)}, U) \cong \operatorname{Ext}^1_R(U, U)^B$ for any set B, we see that condition (ii) of Theorem 5.2.5 also holds. ■

Any finitely generated module over an artin algebra is pure-injective in the sense of the following definition. Here we shall see that, according to a recent result of S. Bazzoni [7] (Theorem 5.2.12 below), so are all cotilting modules.

Definition 5.2.8. A short exact sequence in Mod-R is said to be *pure-exact* if $\operatorname{Hom}_R(M, \_)$ preserves its exactness for every finitely presented module M. A module U_R is *pure-injective* if $\operatorname{Hom}_R(\_, U)$ preserves the exactness of every pure-exact sequence. (For details, see [52].)

To reach our present objective of proving that all cotilting modules are pure-injective, we need the following characterization of pure-injectivity.

Proposition 5.2.9. *A module* U_R *is pure-injective if and only if, for every set* A, $\operatorname{Hom}_R(\_, U)$ *preserves the exactness of the canonical short exact sequence*

$$0 \to U^{(A)} \to U^A \to U^A/U^{(A)} \to 0.$$

Proof. See [52, Theorem 7.1]. ∎

Thus it is straightforward that, if U_R is a cotilting module, then U_R is pure-injective if and only if $\operatorname{Ext}^1_R(U^A/U^{(A)}, U) = 0$ (equivalently, $U^A/U^{(A)}$ is U-torsionless) for every set A. The following simple lemma reduces the task of proving that this condition holds for all cotilting modules to the countable case.

Lemma 5.2.10. *If* $U^{\mathbb{N}}/U^{(\mathbb{N})} \in \operatorname{Cogen}(U_R)$, *then* $U^A/U^{(A)} \in \operatorname{Cogen}(U_R)$ *for all sets* A.

Proof. Under the hypothesis it clearly suffices to show that the $U^A/U^{(A)} \in \operatorname{Cogen}(U^{\mathbb{N}}/U^{(\mathbb{N})})$. So assume that $x = (y_\alpha)_{\alpha \in A} + U^{(A)}$ is a non-zero element of $U^A/U^{(A)}$. Then there is a countably infinite subset $B \subseteq A$ such that $y_\alpha \neq 0$ for all $\alpha \in B$, and so the the kernel of the canonical mapping $U^A/U^{(A)} \to U^B/U^{(B)} \cong U^{\mathbb{N}}/U^{(\mathbb{N})}$ does not contain x. ∎

We need two results from set theory. A family of sets is *almost disjoint* if the intersection of any two distinct elements of the family is finite.

Lemma 5.2.11. *(1) For any infinite set* A *there is a family of* $\operatorname{card}(A^{\mathbb{N}})$ *countable almost disjoint subsets of* A.
(2) For any cardinal ν *there is a cardinal* $\lambda \geq \nu$ *such that* $\lambda^{\aleph_0} = 2^\lambda$.

Proof. (1) Let $T = \{t : \{1, \ldots, n\} \to A \mid n \in \mathbb{N}\}$. Then we have $\operatorname{card}(T) = \operatorname{card}(\{F \subseteq A \mid F \text{ is finite}\}) \cdot \aleph_0 = \operatorname{card}(A)$. For every function $f \in A^{\mathbb{N}}$, let $T_f = \{f|_{\{1,\ldots,n\}} \mid n \in \mathbb{N}\}$. If f and g are two different functions in $A^{\mathbb{N}}$, then $T_f \cap T_g$ is finite. Indeed, if m is the least element of $\mathbb{N}$ such that $f(m) \neq g(m)$, then $T_f \cap T_g = \{f|_{\{1,\ldots,n\}} \mid n < m\}$ has just $m - 1$ elements. Thus $\{T_f\}_{f \in A^{\mathbb{N}}}$ is a family of $\operatorname{card}(A^{\mathbb{N}})$ countable almost disjoint subsets of T. Considering a bijection of A to T, we have the asserted conclusion.
(2) We reproduce the proof given in [43, Lemma 3.2]. Let $\alpha_1 = \nu$ and for $n \geq 1$ define $\alpha_{n+1} = 2^{\alpha_n}$. Let

$$\lambda = \sum_{n \in \mathbb{N}} \alpha_n \geq \nu.$$

Then

$$\lambda^{\aleph_0} \geq \prod_{n \in \mathbb{N}} \alpha_n = \prod_{n \in \mathbb{N}} 2^{\alpha_{n+1}} = 2^{\sum_{n \in \mathbb{N}} \alpha_{n+1}} = 2^{\sum_{n \in \mathbb{N}} \alpha_n} = 2^\lambda.$$

Since clearly $\lambda^{\aleph_0} \leq 2^\lambda$, the conclusion follows. ■

Now we obtain Bazzoni's theorem.

Theorem 5.2.12. *If U_R is a cotilting module, then U_R is pure-injective.*

Proof. According to Lemma 5.2.10 and the remarks preceding it, it suffices to prove that $\operatorname{Ext}^1_R(U^{\mathbb{N}}/U^{(\mathbb{N})}, U_R) = 0$. Let $X = U^{\mathbb{N}}/U^{(\mathbb{N})}$. Consider an infinite set A and, applying Lemma 5.2.11(1), let $\{A_\beta\}_{\beta \in A^{\mathbb{N}}}$ be an almost disjoint family of card($A^{\mathbb{N}}$) countable subsets of A. For each $\beta \in A^{\mathbb{N}}$, let $\eta_\beta : U^{A_\beta} \to U^A/U^{(A)}$ be the restriction to U^{A_β} of the canonical mapping of U^A onto $U^A/U^{(A)}$. Then $\operatorname{Ker} \eta_\beta = U^{(A_\beta)}$ and $\operatorname{Im} \eta_\beta = X_\beta \cong X$.

We now claim that, since the family $\{A_\beta\}$ is almost disjoint, the sum $\Sigma_{A^{\mathbb{N}}} X_\beta \leq U^A/U^{(A)}$ is a direct sum. For suppose $\Sigma_{i=1}^n \eta_{\beta_i}(z_{\beta_i}) = 0$ for some $z_{\beta_i} \in U^{A_{\beta_i}}$. Then $y = \Sigma_{i=1}^n z_{\beta_i} \in U^{(A)}$. Let F be the support of y and $G = \cup_{1 \leq i \neq j \leq n}(A_{\beta_i} \cap A_{\beta_j})$. Then $F \cup G$ is finite. Fix an index $i \in \{1, \dots, n\}$. For every $\alpha \in A_{\beta_i} \backslash (F \cup G)$, the α-component $y(\alpha)$ of y is 0, and hence $z_{\beta_i}(\alpha) = 0$ too since $\alpha \notin A_{\beta_j}$ whenever $j \neq i$. Thus $z_{\beta_i} \in U^{(A)}$, and hence $\eta_{\beta_i}(z_{\beta_i}) = 0$ for every $i \in \{1, \dots, n\}$, proving our claim.

Let V be the submodule of U^A with $V/U^{(A)} = \oplus_{A^{\mathbb{N}}} X_\beta$. By Lemma 5.2.11(2), we can choose A such that $\lambda = \operatorname{card}(A) \geq \operatorname{card}(\Delta U)$ and $\lambda^{\aleph_0} = 2^\lambda$. Since $\operatorname{Ker}(\Gamma) = \operatorname{Cogen}(U_R)$ is closed under submodules and direct products, we have $\Gamma(V) = 0$, so the exact sequence

$$0 \to U^{(A)} \longrightarrow V \longrightarrow X^{(A^{\mathbb{N}})} \to 0$$

induces an exact sequence

$$\Delta(U^{(A)}) \longrightarrow \Gamma(X^{(A^{\mathbb{N}})}) \to 0.$$

The first term has $\operatorname{card}((\Delta U)^A) \leq \lambda^\lambda = 2^\lambda$. The second term is isomorphic to $\Gamma(X)^{A^{\mathbb{N}}}$. Thus, if $\Gamma(X) \neq 0$, the second term has cardinality at least $2^{\lambda^{\aleph_0}} = 2^{2^\lambda}$, contradicting the existence of this epimorphism. ■

Now we have the following result, which appeared in [59] and will prove useful in the following section.

Corollary 5.2.13. *If U_R is a pure-injective module, then Ker Γ is closed under direct limits. In particular, if U_R is a cotilting module, then direct limits of U-torsionless modules are U-torsionless.*

Proof. Consider a direct system $\{M_i\}_{i \in I}$ in Ker(Γ). According to [80, 33.9(2)], the canonical exact sequence

$$0 \to K \xrightarrow{f} \oplus_{i \in I} M_i \to \varinjlim M_i \to 0$$

is pure-exact, so since U_R is pure-injective, Δf is surjective in the induced exact sequence

$$\Delta(\oplus_{i \in I} M_i) \xrightarrow{\Delta f} \Delta(K) \to \Gamma(\varinjlim M_i) \to \Gamma(\oplus_{i \in I} M_i) = 0$$

and the first statement follows. The final statement follows from Theorem 5.2.12. ∎

5.3. Cotilting Bimodules

According to Propositions 3.2.2 and 5.2.7, if U_R is a finitely generated cotilting module over an artin algebra R and $S = \text{End}(U_R)$, then $_S U_R$ is faithfully balanced and $_S U$ is a finitely generated cotilting module. Thus we are led to the following

Definition 5.3.1. A *cotilting bimodule* is a faithfully balanced bimodule $_S U_R$ such that U_R and $_S U$ are cotilting modules.

Remark 5.3.2. Since, according to Proposition 5.2.3, cotilting modules are costar modules, if $_S U_R$ is a cotilting bimodule, then all finitely generated U-torsionless right R-modules and all finitely generated U-torsionless left S-modules are U-reflexive.

We shall employ Proposition 5.1.6 to obtain the cotilting theorem of R. Colpi [26] induced by a cotilting bimodule. To do so we need the following lemmas.

Lemma 5.3.3. *Let $_R U_S$ be a cotilting bimodule and let $0 \to K \longrightarrow M \longrightarrow L \to 0$ be exact. Then*

(1) if $L \in \text{Ker} \, \Gamma$ and any two of K, M, L are reflexive, then so is the third;

(2) if $K, L \in \operatorname{Ker} \Gamma$, then M is U-reflexive if and only if K, L are U-reflexive.

Proof. Since, in either case $L \in \operatorname{Ker} \Gamma$, and $\Gamma \Delta = 0$ when $_S U_R$ is a cotilting bimodule, we have a commutative diagram

$$
\begin{array}{ccccccccc}
0 \to & K & \to & M & \to & L & \to 0 \\
& \delta_K \downarrow & & \delta_M \downarrow & & \delta_L \downarrow & \\
0 \to & \Delta^2 K & \to & \Delta^2 M & \to & \Delta^2 L & \to 0
\end{array}
$$

with exact rows. In each case the Five Lemma applies. ∎

The next lemma is derived from the work of F. Mantese in [58].

Lemma 5.3.4. *Let $_R U_S$ be a cotilting bimodule, let $\mathcal{F}$ denote the class of U-reflexive modules in* Mod-R, *and let*

$$
\mathcal{A} = \{ M_1 / M_2 \mid M_i \in \mathcal{F} \text{ for } i = 1, 2 \}.
$$

Suppose that $0 \to K \longrightarrow M \longrightarrow L \to 0$ is exact in Mod-R. *Then*

(1) if $M \in \mathcal{F}$ and $L \in \mathcal{A}$, then $K \in \mathcal{F}$;
(2) if $K \in \mathcal{F}$, then $M \in \mathcal{A}$ if and only if $L \in \mathcal{A}$;
(3) if $M \in \mathcal{A}$, then $K \in \mathcal{A}$ if and only if $L \in \mathcal{A}$;
(4) if $M \in \mathcal{A}$ and $L \in \operatorname{Cogen}(U)$, then $L \in \mathcal{F}$.

Proof. Let $0 \to L_2 \longrightarrow L_1 \longrightarrow L \to 0$ be exact with $L_i \in \mathcal{F}$. If $M \in \mathcal{F}$, then in the pullback diagram

$$
\begin{array}{ccc}
& 0 & 0 \\
& \uparrow & \uparrow \\
0 \to K \longrightarrow & M \longrightarrow & L \to 0 \\
\| & \uparrow & \uparrow \\
0 \to K \longrightarrow & P \longrightarrow & L_1 \to 0, \\
& \uparrow & \uparrow \\
& L_2 = & L_2 \\
& \uparrow & \uparrow \\
& 0 & 0
\end{array}
$$

$P \in \mathcal{F}$ by the first column and Lemma 5.3.3(2), so $K \in \mathcal{F}$ by Lemma 5.3.3(2) and the second row. Thus (1) follows. Also if $K \in \mathcal{F}$, then $P \in \mathcal{F}$ by Lemma 5.3.3(2), and so $M \in \mathcal{A}$ by the first column, proving one implication of (2).

For the remainder of the proof we consider the commutative diagram with exact rows and columns and the $M_i \in \mathcal{F}$

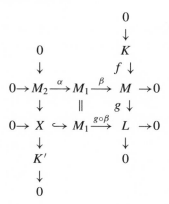

where $K \cong K'$ by the Snake Lemma. If $K \in \mathcal{F}$, then $X \in \mathcal{F}$ by Lemma 5.3.3(2) applied to the first column. Thus the second row shows that $L \in \mathcal{A}$. Thus (2) holds.

To verify (3), from the first column, if $K \in \mathcal{A}$ it follows from (2) that $X \in \mathcal{A}$, so since $X \in \operatorname{Ker} \Gamma$, $X \in \mathcal{F}$ by Lemma 5.3.3(2). Thus by (2) and the second row, $L \in \mathcal{A}$. Also from the second row, if $L \in \mathcal{A}$, then $X \in \mathcal{F}$ by (1), so $K \cong K' \in \mathcal{A}$.

Finally, to verify (4), if $L \in \operatorname{Cogen}(U_R)$, then $L \in \mathcal{F}$ by Lemma 5.3.3(2) applied to the second row. ∎

These last two lemmas provide the tools to prove that the category of R-maps with objects $\mathcal{A}$, as in Lemma 5.3.4, is an abelian subcategory of Mod-R. We also need

Lemma 5.3.5. *Let $_S U_R$ be a bimodule and let Δ denote the U-dual.*

(1) If $\Gamma \Delta^2 M = 0$, then $\operatorname{Coker}(\delta_M) \in \operatorname{Ker} \Gamma$.
(2) If ΔM is U-reflexive, then $\operatorname{Coker}(\delta_M) \in \operatorname{Ker} \Delta$.

Proof. Since $\Delta M \cong \Delta(M / \operatorname{Rej}_U(M))$, the exact sequence

$$0 \to M / \operatorname{Rej}_U(M) \longrightarrow \Delta^2 M \longrightarrow \operatorname{Coker} \delta_M \to 0$$

yields, under an application of Δ, an exact sequence

$$0 \to \Delta(\operatorname{Coker} \delta_M) \longrightarrow \Delta^3 M \xrightarrow{\Delta(\delta_M)} \Delta M \to \Gamma(\operatorname{Coker} \delta_M) \to 0$$

if $\Delta^2 M \in \text{Ker }\Gamma$. But $\Delta(\delta_M)$ is epic since $\Delta(\delta_M) \circ \delta_{\Delta M} = 1_{\Delta M}$, so (1) holds. If $\delta_{\Delta M}$ is an isomorphism, then so is $\Delta(\delta_M)$ and (2) follows. ∎

Note that the abelian subcategories $\mathcal{A}_R$ and $_S\mathcal{A}$ of Mod-R and S-Mod in a cotilting theorem, contain the finitely presented modules in Mod-R and S-Mod, respectively.

Theorem 5.3.6. *Let $_SU_R$ be a cotilting bimodule. Let $\mathcal{F}_R$ and $_S\mathcal{F}$ denote the classes of U-reflexive modules in Mod-R and S-Mod, respectively, and let*

$$\mathcal{A}_R = \{M_1/M_2 \mid M_1, M_2 \in \mathcal{F}_R\} \quad and \quad _S\mathcal{A} = \{N_1/N_2 \mid N_1, N_2 \in {}_S\mathcal{F}\}.$$

Then $_SU_R$ induces a cotilting theorem (with classes of torsion free objects $\mathcal{F}_R$ and $_S\mathcal{F}$) between $\mathcal{A}_R$ and $_S\mathcal{A}$.

Proof. We begin by showing that $\mathcal{A}_R$ is an abelian subcategory of Mod-R. (This, too, employs the work of Mantese [58].) Clearly $\mathcal{A}_R$ is closed under finite direct sums. So suppose that $f : M \to M'$ in $\mathcal{A}_R$ and consider the pushout diagram where $M' \cong M_1'/M_2'$ with the $M_i' \in \mathcal{F}_R$.

$$
\begin{array}{ccc}
 & 0 & \quad 0 \\
 & \uparrow & \quad \uparrow \\
0 \to \text{Ker }f \to M & \xrightarrow{f} & M' \\
\quad\;\| & \uparrow & n \uparrow \\
0 \to \text{Ker }f \to P & \xrightarrow{f'} & M_1' \\
 & \uparrow & \quad \uparrow \\
 & M_2' & = \quad M_2' \\
 & \uparrow & \quad \uparrow \\
 & 0 & \quad 0.
\end{array}
$$

Applying Lemma 5.3.4(2) to the left column, we see that $P \in \mathcal{A}$. Then Im $f' \in \mathcal{F}_R$ by Lemma 5.3.4(4), and so Ker $f \cong$ Ker $f' \in \mathcal{A}_R$ by Lemma 5.3.4(3). Finally, applying Lemma 5.3.4(3) to the exact sequences

$$0 \to \text{Ker }f \to M \to \text{Im }f \to 0 \quad and \quad 0 \to \text{Im }f \to M' \to \text{Coker }f \to 0,$$

we see that $\mathcal{A}_R$ and, similarly, $_S\mathcal{A}$ are abelian subcategories of Mod-R and S-Mod, respectively.

Clearly $\mathcal{F}_R \subseteq \text{Ker }\Gamma \cap \mathcal{A}_R$. But if $M \in \text{Ker }\Gamma = \text{Cogen}(U_R)$ and

$$0 \to M_2 \longrightarrow M_1 \longrightarrow M \to 0$$

is exact with the $M_i \in \mathcal{F}_R$, then by Lemma 5.3.3(1), $M \in \mathcal{F}_R$. Thus $\mathcal{F}_R =$ Ker $\Gamma \cap \mathcal{A}_R$. By definition $\mathcal{T}_R =$ Ker $\Delta \cap \mathcal{A}_R$.

Next we verify condition (2) of Definition 5.1.1. From an exact sequence

$$0 \to M_2 \xrightarrow{f} M_1 \xrightarrow{g} M \to 0$$

with the $M_i \in \mathcal{F}_R$, we obtain an exact sequence

$$0 \to \Delta M \xrightarrow{\Delta g} \Delta M_1 \xrightarrow{\Delta f} \Delta M_2 \xrightarrow{\partial} \Gamma M \to 0 \qquad \text{(ex-1)}$$

from which, letting $I = \text{Im } \Delta f$, we obtain two short exact sequences

$$0 \to \Delta M \xrightarrow{\Delta g} \Delta M_1 \xrightarrow{\alpha} I \to 0 \qquad \text{(ex-a)}$$

and

$$0 \to I \xrightarrow{\beta} \Delta M_2 \xrightarrow{\partial} \Gamma M \to 0 \qquad \text{(ex-b)}$$

where $\Delta f = \beta \circ \alpha$. In ex-a, I and ΔM are torsionless and ΔM_1 is U-reflexive. Thus, by Lemma 5.3.3, ΔM and I are reflexive. Therefore

$$\Delta : \mathcal{A}_R \to {}_S\mathcal{F},$$

and by ex-b, $\Gamma M \in {}_S\mathcal{A}$. Applying Δ to ex-1, we obtain a commutative diagram

$$
\begin{array}{ccccc}
0 & \to & M_2 & \xrightarrow{f} & M_1 \\
 & & \delta_{M_2} \downarrow & & \delta_{M_1} \downarrow \\
0 \to \Delta\Gamma M & \xrightarrow{\Delta\partial} & \Delta^2 M_2 & \xrightarrow{\Delta^2 f} & \Delta^2 M_1
\end{array}
$$

from which, since the M_i are U-reflexive, we see that

$$\Gamma : \mathcal{A}_R \to {}_S\mathcal{A} \cap \text{Ker } \Delta = {}_S\mathcal{T}.$$

Now we turn to condition (1) of Definition 5.1.1. Since $U_R \in \mathcal{F}_R \subseteq \text{Cogen}(U_R)$,

$$\mathcal{T}_R = \{T \in \mathcal{A}_R \mid \text{Hom}_R(T, F) = 0 \text{ for all } F \in \mathcal{F}\}.$$

Since $\Gamma\Delta^2 = 0$ and ΔM is U-reflexive whenever $M \in \mathcal{A}_R$, we see from Lemma 5.3.5 and Proposition 5.2.6 that δ_M is epic for all $M \in \mathcal{A}_R$. But then the exact sequence

$$0 \to \text{Ker } \delta_M \longrightarrow M \xrightarrow{\delta_M} \Delta^2 M \to 0$$

shows that Ker $\delta_M \in \mathcal{A}_R$, and applying Δ, we obtain an exact sequence

$$0 \longrightarrow \Delta^3 M \xrightarrow{\Delta\delta_M} \Delta M \to \Delta(\text{Ker } \delta_M) \to 0 = \Gamma\Delta^2 M,$$

which, since $\Delta\delta_M$ is an epimorphism, shows that $\Delta(\mathrm{Ker}\,\delta_M) = 0$, that is, $\mathrm{Ker}\,\delta_M \in \mathcal{T}_R$. Thus if $M \in \mathcal{A}_R\setminus\mathcal{F}_R$ there is a $T = \mathrm{Ker}\,\delta_M \in \mathcal{T}_R$ such that $\mathrm{Hom}_R(T, M) \neq 0$. So we see that $(\mathcal{T}_R, \mathcal{F}_R)$ is a torsion theory in $\mathcal{A}_R$ with $\tau(M) = \mathrm{Ker}\,\delta_M$.

Clearly the first part of condition (4) of Definition 5.1.1 and the remaining conditions of Proposition 5.1.6 are satisfied, and so the theorem is proved. ∎

Since, according to Corollary 5.2.13, if $_SU_R$ is a cotilting bimodule, $\mathrm{Cogen}(U_R) = \mathrm{Ker}\,\Gamma_{U_R}$ and $\mathrm{Cogen}(_SU) = \mathrm{Ker}\,\Gamma_{_SU}$ are closed under direct limits, using a proof similar to Proposition 4.3.18, we have the following characterization of the U-reflexive modules that is a consequence of results in [7] and [59].

Proposition 5.3.7. *If $_SU_R$ is a cotilting bimodule, then M_R is U-reflexive if and only if M is U-dense and U-linearly compact.*

Proof. In view of Lemma 4.3.15, we only need to prove necessity. So assume that M is U-reflexive. Then M is trivially U-dense. So suppose that

$$0 \to K_\lambda \xrightarrow{\iota_\lambda} M \xrightarrow{p_\lambda} L_\lambda \to 0$$

is an inverse system of exact sequence with M reflexive and $L_\lambda \in \mathrm{Cogen}(U_R) = \mathrm{Ker}\,\Gamma$. Then, according to Lemma 5.3.3, each K_λ and each L_λ is reflexive. An application of Δ yields a direct system of exact sequences

$$0 \to \Delta L_\lambda \xrightarrow{\Delta p_\lambda} \Delta M \xrightarrow{\Delta \iota_\lambda} \Delta K_\lambda \to 0$$

so that

$$0 \to \varinjlim \Delta L_\lambda \xrightarrow{\varinjlim \Delta p_\lambda} \Delta M \xrightarrow{\varinjlim \Delta \iota_\lambda} \varinjlim \Delta K_\lambda \to 0$$

is exact. Now, since by Corollary 5.2.13 $\varinjlim \Delta K_\lambda \in \mathrm{Cogen}(_SU) = \mathrm{Ker}\,\Gamma$, we have a commutative diagram with exact rows

$$
\begin{array}{ccccccccc}
0 & \to & \Delta(\varinjlim\Delta K_\lambda) & \xrightarrow{\Delta(\varinjlim\Delta\iota_\lambda)} & \Delta^2 M & \xrightarrow{\Delta(\varinjlim\Delta p_\lambda)} & \Delta(\varinjlim\Delta L_\lambda) & \to & 0 \\
 & & \cong \uparrow & & \| & & \cong \uparrow & & \\
0 & \to & \varinjlim\Delta^2 K_\lambda & \xrightarrow{\varinjlim\Delta^2\iota_\lambda} & \Delta^2 M & \xrightarrow{\varinjlim\Delta^2 p_\lambda} & \varinjlim\Delta^2 L_\lambda & & \\
 & & \cong \uparrow & & \cong \uparrow & & \cong \uparrow & & \\
0 & \to & \varinjlim K_\lambda & \xrightarrow{\varinjlim\iota_\lambda} & M & \xrightarrow{\varinjlim p_\lambda} & \varinjlim L_\lambda & &
\end{array}
$$

so that $\varinjlim p_\lambda$ is an epimorphism. ∎

5.4. Cotilting via Tilting and Morita Duality

Here, following [29], we present examples of cotilting bimodules over noetherian serial rings that are not finitely generated. To do so we first consider when the Morita dual of a tilting bimodule is a cotilting bimodule.

Suppose that $_A V_S$ is a tilting bimodule, and let

$$H = \mathrm{Hom}_A(V, \_), \quad H' = \mathrm{Ext}_A^1(V, \_), \quad T = (V \otimes_S \_), \quad T' = \mathrm{Tor}_1^S(V, \_)$$

so that

$$H : \mathrm{Ker}\, H' \rightleftarrows \mathrm{Ker}\, T' : T \quad \text{and} \quad H' : \mathrm{Ker}\, H \rightleftarrows \mathrm{Ker}\, T : T'$$

are pairs of category equivalences, as in the Tilting Theorem 3.5.1. Suppose further that a bimodule $_A W_R$ induces a Morita duality

$$\Delta_W = \Delta_{W_R} : \mathcal{L}_R \rightleftarrows {}_A\mathcal{L} : \Delta_{{}_A W} = \Delta_W$$

between the categories of linearly compact modules in Mod-R and A-Mod (Corollary 4.4.4), and let $_S U_R = \mathrm{Hom}_A(V, W)$, so that

$$_S U_R = \Delta_{{}_A W}({}_A V_S) = H({}_A W_R).$$

Then $_S U$ is a cotilting module according to Proposition 2.3.5, and noting that $_A V \in {}_A\mathcal{L}$ and $_A W \in \mathrm{Ker}\, H'$, we see that $\mathrm{End}(U_R) \cong \mathrm{End}({}_A V) \cong S$, canonically, ([1, Proposition 23.3]) and $\mathrm{End}(_S U) \cong \mathrm{End}({}_A W) \cong R$, canonically ([1, Proposition 21.2]). Thus $_S U_R$ is a faithfully balanced bimodule such that $_S U$ is a cotilting module.

We shall show that in some cases $_S U_R$ is actually a cotilting bimodule, that is, that U_R is a cotilting module. But first we investigate the properties of U_R in this general case.

Using the usual covariant and contravariant adjointness conditions [1, Propositions 20.7 and 20.6], one easily checks that there are natural isomorphisms

$$\Delta_{U_R} \cong H \circ \Delta_{W_R} : \mathrm{Mod}\text{-}R \longrightarrow S\text{-Mod} \tag{delta-1}$$

and

$$\Delta_{_S U} \cong \Delta_{{}_A W} \circ T : S\text{-Mod} \longrightarrow \mathrm{Mod}\text{-}R. \tag{delta-2}$$

To obtain analogous descriptions of the Γ's we need the case $n = 1$ of the following lemma in which the first isomorphism is [11, Proposition 5.1, page 120], and the second can easily be obtained by applying

$$\mathrm{Hom}_A(\_, \mathrm{Hom}_R(M, W)) \cong \mathrm{Hom}_R(M, \mathrm{Hom}_A(\_, W))$$

to a projective resolution of $_A V$.

Lemma 5.4.1. *Let $_SN$ and M_R be modules and $_AV_S$ and $_AW_R$ be bimodules.*

(1) If $_AW$ is injective, then there are natural isomorphisms

$$\mathrm{Hom}_A(\mathrm{Tor}_n^S(V, N), W) \cong \mathrm{Ext}_S^n(N, \mathrm{Hom}_A(V, W))$$

for $n = 1, 2, \ldots$.

(2) If $_AW$ and W_R are both injective, then there are natural isomorphisms

$$\mathrm{Ext}_A^n(V, \mathrm{Hom}_R(M, W)) \cong \mathrm{Ext}_R^n(M, \mathrm{Hom}_A(V, W))$$

for $n = 1, 2, \ldots$.

Now we see that there are also natural isomorphisms

$$\Gamma_{U_R} \cong H' \circ \Delta_{W_R} : \mathrm{Mod}\text{-}R \longrightarrow S\text{-}\mathrm{Mod} \qquad\qquad \text{(gamma-1)}$$

and

$$\Gamma_{_SU} \cong \Delta_{_AW} \circ T' : S\text{-}\mathrm{Mod} \longrightarrow \mathrm{Mod}\text{-}R. \qquad\qquad \text{(gamma-2)}$$

As we noted above, $_SU$ is a cotilting module when $_SU_R$ is the Morita dual of a tilting bimodule. Next, we observe that $_SU_R$ is nearly a cotilting bimodule.

Proposition 5.4.2. *Suppose that $_AW_R$ induces a Morita duality and $_AV_S$ is a tilting bimodule. If $_SU_R = \mathrm{Hom}_A(V, W)$, then $_SU_R$ is a faithfully balanced bimodule such that $_SU$ is a cotilting module, and U_R satisfies*

(1) There is an exact sequence $0 \to U_R \longrightarrow W_0 \longrightarrow W_1 \to 0$ with $W_0, W_1 \in \mathrm{add}(W_R)$. In particular, $\mathrm{inj.dim.}U_R \leq 1$ and U_R is finitely cogenerated and linearly compact;

(2) $\Gamma_{U_R}(M) = 0$ whenever $M_R \hookrightarrow U^n$, for $n = 1, 2, \ldots$. In particular, $\mathrm{Ext}_R^1(U, U) = 0$;

(3) There is an exact sequence $0 \to U_1 \longrightarrow U_0 \longrightarrow W_R \to 0$ with $U_0, U_1 \in \mathrm{add}(U_R)$.

Proof. (1) and (3) follow from applications of $\Delta_{_AW}$ to the exact sequences $0 \to P_1 \longrightarrow P_0 \longrightarrow {_AV} \to 0$ with $P_0, P_1 \in \mathrm{add}(_AA)$ and $0 \to {_AA} \longrightarrow V_0 \longrightarrow V_1 \to 0$ with $V_0, V_1 \in \mathrm{add}(_AV)$ that exist because $_AV$ is a tilting module. According to (1), $\mathrm{inj.dim.}U_R \leq 1$, so Γ_{U_R} is right exact. Thus if $M_R \hookrightarrow U^n$, there is an exact sequence $\Gamma_{U_R}(U)^n \to \Gamma_{U_R}(M) \to 0$. But $\Gamma_{U_R}(U) \cong H' \circ \Delta_{W_R}(\Delta_{_AW}(V)) \cong H'(V) = 0$. ∎

As we have just seen, the module U_R that we are considering satisfies conditions (i) and (iii) of the characterization of cotilting modules in Theorem 5.2.5. Thus U_R is a cotilting module when condition (ii) of that theorem holds.

Corollary 5.4.3. *If $_S U_R$ is as in Proposition 5.4.2, then $_S U_R$ is a cotilting bimodule if and only if $\mathrm{Ext}_R^1(U^A, U) = 0$ for all sets A.*

Suppose R is a noetherian serial ring. Then either R is artinian, R has no artinian direct summands, or R is a ring direct sum $R = R_1 + R_2$ such that R_1 has no artinian ring direct summands and R_2 is artinian. Since a module U_R is a cotilting module if and only if the component of U belonging to each each indecomposable ring direct summand of R is a cotilting module, to show that U_R is a cotilting module, we may assume that R is indecomposable and either artinian or not artinian. We shall show that if R has self-duality induced by a bimodule $_R W_R$ and $_R V_S$ is a tilting bimodule, then $_S U_R = \Delta_W(V)$ is a cotilting bimodule. In view of the preceding discussion, to do so we need only show that $\mathrm{Ext}_R^1(U^A, U) = 0$ for all sets A, which is a fact that will become apparent from

Lemma 5.4.4. *If U_R is a finitely cogenerated module over a noetherian serial ring such that $\mathrm{Ext}_R^1(U, U) = 0$, then $\mathrm{Ext}_R^1(U^A, U) = 0$ for any set A.*

Proof. Assume that R is indecomposable and $\mathrm{Ext}_R^1(U, U) = 0$. If R is artinian, then U_R is finitely generated and, according to Propositions B.1.6 and B.1.7, if $S = \mathrm{End}(U_R)$, then S is artinian and $_S U \cong \mathrm{Hom}_R(R, U)$ is finitely generated. Thus by Proposition 1.3.2 $U^A \epsilon$ Add(U). But then, $\mathrm{Ext}_R^1(U^A, U) = 0$.

If R is not artinian, then we may assume that

$$U_R = E_{i_1} \oplus \cdots \oplus E_{i_k} \oplus M_1 \oplus \cdots \oplus M_\ell$$

where the E_{i_j} are injective envelopes of their socles and the M_i are of finite length. Then we see from Proposition B.2.2 and the argument in the first paragraph of this proof that

$$U^A \cong E_{i_1}^{(B_1)} \oplus \cdots \oplus E_{i_k}^{(B_k)} \oplus E_0^{(C)} \oplus M_1^{(D_1)} \oplus \cdots \oplus M_\ell^{(D_\ell)}$$

for some sets B_j, C, D_i. Now to see that $\mathrm{Ext}_R^1(U^A, U) = 0$ we need only check that each $\mathrm{Ext}_R^1(E_0, M_i) = 0$. To do so, observe that, by Proposition

B.2.1, there is an injective resolution

$$0 \to M_i \longrightarrow E_j \xrightarrow{f} E_k \to 0$$

with E_j and E_k indecomposable and artinian. Then we need to show that

$$\mathrm{Hom}_R(E_0, E_j) \xrightarrow{\mathrm{Hom}_R(E_0, f)} \mathrm{Hom}_R(E_0, E_k) \to 0$$

is exact. So, let $\beta \in \mathrm{Hom}_R(E_0, E_k)$ with $K = \mathrm{Ker}\,\beta$. Then by Proposition B.2.1 there is an $m \in \mathbb{N}$ such that $E_0/KJ^m \cong E_{i_1}$, and since $\mathrm{Ext}^1_R(E_{i_1}, M_i) = 0$

$$\mathrm{Hom}_R(E_0/KJ^m, E_j) \xrightarrow{\mathrm{Hom}_R(E_0/KJ^m, f)} \mathrm{Hom}_R(E_0/KJ^m, E_k) \to 0$$

is exact. Thus, letting $\eta : E_0 \to E_0/KJ^m$ be the natural map, we have a map $\gamma \in \mathrm{Hom}_R(E_0/KJ^m, E_j)$ and a commutative diagram

$$
\begin{array}{ccc}
E_0 & \xrightarrow{\beta} & E_k \\
\eta \downarrow & & \| \\
E_0/KJ^m & \xrightarrow{\overline{\beta}} & E_k \\
\gamma \downarrow & & \| \\
E_j & \xrightarrow{f} & E_k
\end{array}
$$

so that $\mathrm{Hom}_R(E_0, f)(\gamma \circ \eta) = f \circ \gamma \circ \eta = \overline{\beta} \circ \eta = \beta$. ∎

Now, as promised, we have

Proposition 5.4.5. *Let R be a noetherian serial ring with self-duality induced by a duality bimodule ${}_R W_R$. If ${}_R V_S$ is a tilting bimodule, then ${}_S U_R = \Delta_W(V)$ is a cotilting bimodule.*

Proof. Since U_R is finitely cogenerated and $\mathrm{Ext}^1_R(U, U) = 0$ by Proposition 5.4.2, Lemma 5.4.4 and Corollary 5.4.3 complete the proof. ∎

Non-artinian noetherian serial rings with self-duality are rather abundant (see Appendix B), and so we can obtain examples of non-finitely generated cotilting modules. Indeed, if in Example 3.7.12 we insist that $D = K[[x]]$ (or any other linearly compact noetherian uniserial ring), let ${}_R V_S$ be the given tilting module and let ${}_R W$ be the minimal cogenerator, then ${}_R W_R$ induces a self-duality (Proposition B.2.3), and so ${}_S U_R = \Delta_W(V)$ is a cotilting bimodule that is artinian but not finitely generated. Perhaps a simpler example will be edifying.

Example 5.4.6. *Let D be a linearly compact local noetherian serial ring with maximal ideal* M, *let J be the radical of*

$$R = \begin{bmatrix} D & D \\ \text{M} & D \end{bmatrix} \text{ with } e_1 = \begin{bmatrix} 1 & 0 \\ 0 & 0 \end{bmatrix} \text{ and } e_2 = \begin{bmatrix} 0 & 0 \\ 0 & 1 \end{bmatrix},$$

and let $S_i = Re_i/Je_i$ and $T_i = e_i R/e_i J$, for $i = 1, 2$. Here, as in Example 3.7.12, if $P_2 = Re_2$, we see that

$$V = P_2 \oplus S_2$$

is a tilting module with, as one may calculate,

$$S = \text{End}(_R V) \cong \begin{bmatrix} D & C \\ 0 & C \end{bmatrix}$$

with $C = D/$M. Now R has self-duality induced by a bimodule $_R W_R$, and by Proposition 5.4.5 $_S U_R = \Delta_W(V)$ is a cotilting bimodule. We can write

$$_R W = Q_1 \oplus Q_2 \text{ and } W_R = E_1 \oplus E_2$$

where $Q_i = E(S_i)$ and $E_i = E(T_i)$. Let us assume that $\Delta_W(P_2) = E_2$ and, consequently, $\Delta_W(S_2) = T_2$. Then

$$U_R \cong E_2 \oplus T_2.$$

On the other hand, letting maps operate on the opposite side of modules,

$$_S U \cong \text{Hom}_R(V, W)$$
$$\cong \begin{bmatrix} \text{Hom}_R(P_2, Q_1) & \text{Hom}_R(P_2, Q_2) \\ \text{Hom}_R(S_2, Q_1) & \text{Hom}_R(S_2, Q_2) \end{bmatrix}$$
$$\cong \begin{bmatrix} I & I \\ 0 & C \end{bmatrix}$$

where, after identifying $D = e_2 Re_2$, we see that $_D I = E(_D C) \cong e_2 Q_1 \cong e_2 Q_2$, the unique artinian uniserial D-module of infinite length. Moreover it is interesting to check that the following diagrams indicate the structure of S and U.

Structure of S :

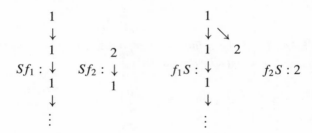

Structure of U :

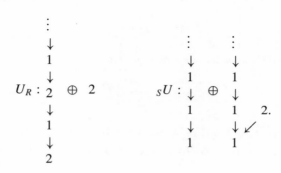

5.5. Weak Morita Duality

As a consequence of Corollary 4.4.4, a Morita duality is a duality between the categories $\mathcal{L}_R$ and $_S\mathcal{L}$ of linearly compact modules in Mod-R and S-Mod with $R_R \in \mathcal{L}_R$ and $_S S \in {}_S\mathcal{L}$. In any case $\mathcal{L}_R$ and $_S\mathcal{L}$ are closed under epimorphic images, submodules, and extensions [81, Proposition 3.3]. Since, when there is a Morita duality between them, $\mathcal{L}_R$ and $_S\mathcal{L}$ are precisely the U-reflexive modules relative to a faithfully balanced module $_S U_R$, we now focus on just the closure properties of $\mathcal{L}_R$ and $_S\mathcal{L}$. We are led to the following more general notion.

Definition 5.5.1. A *weak Morita duality (WMD)* is a duality

$$D_R : \mathcal{F}_R \rightleftarrows {}_S\mathcal{F} : D_S$$

between full subcategories $\mathcal{F}_R$ of Mod-R and $_S\mathcal{F}$ of S-Mod such that

(1) $R_R \in \mathcal{F}_R$ and $_S S \in {}_S\mathcal{F}$;
(2) $\mathcal{F}_R$ and $_S\mathcal{F}$ are closed under submodules and extensions.

According to a theorem of Morita [1, Theorem 23.5], we may assume that $D_R = \mathrm{Hom}_R(\_, U) = \Delta_{U_R}(= \Delta)$ and $D_S = \mathrm{Hom}_S(\_, U) = \Delta_{sU}(= \Delta)$ for a faithfully balanced bimodule $_S U_R$ with $U_R \in \mathcal{F}_R$ and $_S U \in {_S}\mathcal{F}$ and that all modules in $\mathcal{F}_R$ and $_S\mathcal{F}$ are U-reflexive; then we say that $_S U_R$ *induces a weak Morita duality (WMD)* between $\mathcal{F}_R$ and $_S\mathcal{F}$.

A WMD between the subcategories of all U-reflexive modules in Mod-R and S-Mod is called a *generalized Morita duality (GMD)*.

Proposition 5.5.2. *A bimodule $_S U_R$ is a duality module if and only if it induces a weak Morita duality between categories $\mathcal{C}_R$ and $_S\mathcal{C}$, which are closed under epimorphic images and contain R_R and $_S S$, respectively.*

Proof. See [1, Theorem 24.1]. ∎

Note that when a subcategory $\mathcal{F}$ of Mod-R is closed under extensions (and so finite direct sums), $\mathrm{gen}(\mathcal{F})$ is the collection of epimorphic images of modules in $\mathcal{F}$.

Suppose now, for the next two results, that the $_S U_R$-dual Δ induces a WMD between $\mathcal{F}_R$ and $_S\mathcal{F}$, and that

$$0 \to K \xrightarrow{f} M \xrightarrow{g} X \to 0$$

is exact with M (and hence K) $\in \mathcal{F}_R$ (or $_S\mathcal{F}$). Then, according to (1) of the following proposition, the sequence

(#) $$0 \to \Delta X \xrightarrow{\Delta g} \Delta M \xrightarrow{\Delta f} \Delta K \to \Gamma X \to 0$$

is also exact.

Proposition 5.5.3. *If $_S U_R$ induces a WMD between $\mathcal{F}_R$ and $_S\mathcal{F}$, then*

(1) $\mathrm{Ext}^1_R(M, U) = 0$ *for all* $M \in \mathcal{F}_R$;
(2) $\mathrm{Id}(U_R) \le 1$;
(3) $\mathrm{Ker}\,\Delta \cap \mathrm{Ker}\,\Gamma \cap \mathrm{gen}(\mathcal{F}_R) = 0$.

Proof. (1) Since $_S S$ is U-reflexive, so is $U_R \cong \Delta(_S S)$ and $\Delta(U_R) \cong {_S}S$. Thus if $M \in \mathcal{F}_R$ and $0 \to U \longrightarrow X \longrightarrow M \to 0$ is exact, then by Lemma 4.2.4, so is $0 \to \Delta M \longrightarrow \Delta X \longrightarrow \Delta U \to 0$. But this sequence is split exact, so, since U, X, and M are all U-reflexive, $0 \to U \longrightarrow X \longrightarrow M \to 0$ splits.

(2) If $I \leq R_R$, then according to (1) $\operatorname{Ext}^2_R(R/I, U) = \operatorname{Ext}^1_R(I, U) = 0$. Thus, if

$$0 \to U \longrightarrow E \longrightarrow C \to 0$$

is exact in Mod-R with E injective, then $\operatorname{Ext}^1_R(R/I, C) = \operatorname{Ext}^2_R(R/I, U) = 0$, so C is injective by Baer's Criterion.

For (3), in the exact sequence (#), if $\Delta X = 0 = \Gamma X$, Δf and hence $\Delta^2 f$ are isomorphisms. But then, since δ_M and δ_K are isomorphisms, so is f, that is, $X = 0$. ∎

Throughout the remainder of this section, when a result or proof refers symmetrically to both $\mathcal{F}_R$ and $_S\mathcal{F}$, we shall simply denote both categories by $\mathcal{F}$.

Proposition 5.5.4. *Let $_S U_R$ induce a WMD between $\mathcal{F}_R$ and $_S\mathcal{F}$, and suppose that $X \in \operatorname{gen}(\mathcal{F})$. Then*

(1) $\Gamma \Delta X = 0$;
(2) $\Delta \Gamma X = 0$;
(3) δ_X is epic;
(4) $\Delta(\operatorname{Ker} \delta_X) = 0$;
(5) $\Gamma X = 0$ if and only if $X \in \mathcal{F}$.

Proof. Since Δ is left exact we see, as in (#), that $\Delta : \operatorname{gen}(\mathcal{F}_R) \to _S\mathcal{F}$. Thus (1) follows from Proposition 5.5.3.

For (2), applying Δ to the sequence (#) we obtain an exact sequence

$$0 \to \Delta \Gamma X \longrightarrow \Delta^2 K \xrightarrow{\Delta^2 f} \Delta^2 M$$

in which $\Delta^2 f = \delta_M \circ f \circ \delta_K^{-1}$ is monic, that is, $\Delta \Gamma X = 0$.

Let $I = \operatorname{Im} \Delta f$ in (#). Since $I \subset \Delta K \in \mathcal{F}$, by (1) of Proposition 5.5.3 $\Gamma I = 0$. Thus from the commutative diagram with exact rows

$$
\begin{array}{ccccc}
M & \xrightarrow{g} & X & \to 0 & \\
\cong \downarrow & & \delta_X \downarrow & & \\
\Delta^2 M & \xrightarrow{\Delta^2 g} & \Delta^2 X & \to 0 = \Gamma I &
\end{array}
$$

we see that δ_X is epic and (3) holds.

Now, by (3) we have an exact sequence

$$0 \to \operatorname{Ker} \delta_X \longrightarrow X \xrightarrow{\delta_X} \Delta^2 X \to 0,$$

which, since by (1) $\Gamma \Delta^2 X = 0$, yields an exact sequence

$$0 \to \Delta^3 X \xrightarrow{\Delta \delta_X} \Delta X \longrightarrow \Delta(\mathrm{Ker}\, \delta_X) \to 0,$$

so $\Delta(\mathrm{Ker}\, \delta_X) = 0$ because $\Delta \delta_X \circ \delta_{\Delta X} = 1_{\Delta X}$.

Finally, for (5), if $X \in \mathcal{F}$, then $\Gamma X = \Gamma \Delta^2 X = 0$ by (1). Conversely, suppose $\Gamma X = 0$. Then, applying Γ to

$$0 \to \mathrm{Ker}\, \delta_X \longrightarrow X \xrightarrow{\delta_X} \Delta^2 X \to 0,$$

we obtain, using (1) and Proposition 5.5.3, an exact sequence

$$0 = \Gamma \Delta^2 X \to \Gamma X \longrightarrow \Gamma(\mathrm{Ker}\, \delta_X) \longrightarrow \mathrm{Ext}^2(\Delta^2 X, U) = 0,$$

so $\Gamma(\mathrm{Ker}\, \delta_X) \cong \Gamma X = 0$. But $\mathrm{Ker}\, \delta_X \in \mathrm{gen}\,\mathcal{F}$, and by (4) $\Delta(\mathrm{Ker}\, \delta_X) = 0$ also. Thus by Proposition 5.5.3, $\mathrm{Ker}\, \delta_X = 0$, and so by (3) $X \cong \Delta^2 X \in \mathcal{F}$. ∎

Since they are U-reflexive, the modules in $\mathcal{F}_R$ and $_S\mathcal{F}$ are U-torsionless when $_S U_R$ induces a WMD between them. In fact $\mathcal{F}_R$ and $_S\mathcal{F}$ are precisely the U-torsionless modules in $\mathrm{gen}(\mathcal{F}_R)$ and $\mathrm{gen}(_S\mathcal{F})$.

Corollary 5.5.5. *If $_S U_R$ induces a WMD between $\mathcal{F}_R$ and $_S\mathcal{F}$, then $\mathcal{F}_R$ and $_S\mathcal{F}$ consist of the U-torsionless modules in $\mathrm{gen}(\mathcal{F}_R)$ and $\mathrm{gen}(_S\mathcal{F})$. In particular, the finitely generated U-torsionless left R-modules and right S-modules all belong to $\mathcal{F}_R$ and $_S\mathcal{F}$, respectively.*

Proof. Proposition 5.5.4(3) shows that, if $X \in \mathrm{gen}\,\mathcal{F}$ is U-torsionless, then $X \cong \Delta^2 X \in \mathcal{F}$. ∎

Now we are in position to prove that a bimodule that induces a WMD between $\mathcal{F}_R$ and $_S\mathcal{F}$ induces a cotilting theorem between $\mathrm{gen}(\mathcal{F}_R)$ and $\mathrm{gen}(_S\mathcal{F})$.

Theorem 5.5.6. *Let $_S U_R$ induce a WMD between $\mathcal{F}_R$ and $_S\mathcal{F}$. Then $_S U_R$ induces a cotilting theorem (with classes of torsion free objects $\mathcal{F}_R$ and $_S\mathcal{F}$) between $\mathrm{gen}(\mathcal{F}_R)$ and $\mathrm{gen}(_S\mathcal{F})$.*

Proof. Clearly $\mathrm{gen}(\mathcal{F}_R)$ is an abelian subcategory of Mod-R. According to Proposition 5.5.4 (5), $\mathcal{F}_R = \mathrm{Ker}\,\Gamma \cap \mathrm{gen}(\mathcal{F}_R)$, and, by definition, $\mathcal{T}_R = \mathrm{Ker}\,\Delta \cap \mathrm{gen}(\mathcal{F}_R)$. Now $\mathrm{Hom}_R(T, C) = 0$ for all $T \in \mathcal{T}_R$, $C \in \mathcal{F}_R$, since $\mathcal{F}_R$ consists of U-torsionless modules, and if $\mathrm{Hom}_R(X, C) = 0$, for all $C \in \mathcal{F}_R$, then $\Delta X = 0$. If $X \in \mathrm{gen}(\mathcal{F}_R)$ and $\mathrm{Hom}_R(T, X) = 0$ for all $T \in \mathcal{T}_R$, then

$\text{Ker }\delta_X = 0$ by Proposition 5.5.4 (4), and so $X \in \mathcal{F}_R$ by Corollary 5.5.5. Also by Proposition 5.5.4 (4), $\text{Ker }\delta_M = \text{Rej}_U(M)$ is the torsion submodule of M whenever $M \in \text{gen}(\mathcal{F}_R)$. Since analogous results hold in $\text{gen}(_S\mathcal{F})$, (1) of Definition 5.1.1 is verified. Condition (2) follows from Proposition 5.5.4 (1) and (2). The first part of condition (4) of Definition 5.1.1 holds since $_SU_R$ induces a WMD between $\mathcal{F}_R$ and $_S\mathcal{F}$. According to Proposition 5.5.3, $\text{Ext}_R^2(\_, U) = 0$, and the remaining conditions of Proposition 5.1.6 are obvious. ∎

The preceding results also yield

Proposition 5.5.7. *If $_SU_R$ induces a WMD, then U_R and $_SU$ are costar modules. In fact, if $M \in \mathcal{F}$ and*

$$0 \to K \xrightarrow{\ f\ } M \xrightarrow{\ g\ } X \to 0$$

is exact, then Δf is epic if and only if $X \in \mathcal{F}$.

Proof. From the exact sequence (#) we see that Δf is epic if and only if $\Gamma X = 0$, so the last statement follows from Proposition 5.5.4(5). But by Corollary 5.5.5, this is equivalent to $X \in \text{Cogen}(U)$, and thus the first statement follows from Theorem 4.2.7(e). ∎

In view of Proposition 5.5.7, the next corollary follows from Proposition 4.2.11 and Corollary 2.4.13, according to which a faithful costar module over an artin algebra is a cotilting module.

Corollary 5.5.8. *Let U_R be finitely generated over an artin algebra R. If $_SU_R$ induces a WMD, then U_R is a cotilting module. Conversely, if U_R is a cotilting module with $S = \text{End}(U_R)$, then $_SU_R$ induces a WMD between the categories $\mathcal{F}_R$ and $_S\mathcal{F}$ of finitely generated U-torsionless left R- and right S-modules.*

We shall show that, when R is an artin algebra and $_SU_R$ is a finitely generated cotilting bimodule, then the abelian categories of Theorem 5.3.6 are just the finitely generated modules. Immediately we have

Corollary 5.5.9. *If $_SU_R$ is a finitely generated cotilting bimodule over an artin algebra, then $_SU_R$ induces a cotilting theorem between* mod-R *and* S-mod.

Our next goal is to show that this is the "largest possible" cotilting theorem in the artin algebra case.

Suppose R is an artin algebra with center K and dual D, and let $_RW_R = D(_RR_R)$. Then $_RW_R$ is a balanced two-sided injective cogenerator, and by adjointness there are natural isomorphisms

$$\Delta_W \cong D$$

on both Mod-R and R-Mod. Let U_R be a finitely generated module with $S = \text{End}(U_R)$. Then, as we saw in Proposition 5.2.7, U_R is a cotilting module if and only if there is a tilting bimodule $_RV_S$ with $_SU_R = D(_SV_R)$, and then $_SU_R$ is a cotilting bimodule.

When a bimodule $_SU_R$ induces Δ and Γ dualities

$$\Delta_{U_R} : \mathcal{F}_R \rightleftarrows {}_S\mathcal{F} : \Delta_{sU} \quad \text{and} \quad \Gamma_{U_R} : \mathcal{T}_R \rightleftarrows {}_S\mathcal{T} : \Gamma_{sU},$$

as in a cotilting theorem (see Definition 5.1.1), for convenience, let us say that the modules in $\mathcal{F}_R$ and $_S\mathcal{F}$ are Δ_U-reflexive and the modules in $\mathcal{T}_R$ and $_S\mathcal{T}$ are Γ_U-reflexive. Also, if there are no maps other than 0 from M to U, we say that M is U-torsion.

Theorem 5.5.10. *Let R be an artin algebra and let U_R be a finitely generated cotilting module with $S = \text{End}(U_R)$. If M is either a right R-module or a left S-module, then*

(1) *M is Δ_U-reflexive if and only if M is finitely generated and U-torsionless;*

(2) *M is Γ_U-reflexive if and only if M is finitely generated and U-torsion.*

Proof. Let $_RV_S$ be a tilting module with $_SU_R = D(V)$ and recall from the Tilting Theorem 3.5.1 that the bimodule $_SV_R$ induces pairs of equivalences

$$H : \text{Ker } H' \rightleftarrows \text{Ker } T' : T \quad \text{and} \quad H' : \text{Ker } H \rightleftarrows \text{Ker } T : T'.$$

If M_R is finitely generated and U-torsionless, then $DM \in \text{gen}(_RV) = \text{Ker } H' \cap R$-mod, and so from (delta-1) and (delta-2) on page 103 we have natural isomorphisms

$$\Delta_{sU} \circ \Delta_{U_R}(M) \cong DTHD(M) \cong D^2(M) \cong M.$$

If M_R is U-torsion, then $DM \in \text{Ker } H$, and so from (gamma-1) and (gamma-2) on page 104 we have natural isomorphisms

$$\Gamma_{sU} \circ \Gamma_{U_R} \cong DT'H'D(M) \cong D^2(M) \cong M.$$

Thus, since V_S is also a tilting module we see that the conditions in each part of the theorem are sufficient.

To see that the conditions are each necessary recall that, since R is an artin algebra, the D-reflexive R-modules are just the finitely generated R-modules [1, Theorem 24.8].

(1) According to Corollary 1.3.3, if M_R is U-torsionless, then there is an embedding

$$0 \to M \to U^{(A)}$$

of M into a direct sum of copies of U. But then we have an epimorphism

$$_R V^A \cong D(U_R)^A \to D(M) \to 0,$$

and since $_R V^A \in \mathrm{Gen}(_R V) = \mathrm{Ker}\, H'$ (Proposition 1.2.3) so is $D(M)$ whenever M_R is U-torsionless. Thus for any such M_R we finally have, employing the isomorphisms (delta-1) and (delta-2) on page 103,

$$\Delta_{sU} \circ \Delta_{U_R}(M) \cong DTHD(M) \cong D^2(M).$$

Thus if M_R is Δ_U-reflexive, then M_R is D-reflexive and U-torsionless, or equivalently, M_R is finitely generated and U-torsionless.

(2) From (delta-1) on page 103 we see that $D(M) \in \mathrm{Ker}\, H$ whenever $\Delta_{U_R}(M) = 0$. So from the isomorphisms (gamma-1) and (gamma-2) on page 104 we have, for any $M \in \mathrm{Mod}\text{-}R$,

$$\Gamma_{sU} \circ \Gamma_{U_R}(M) \cong DT'H'D(M) \cong D^2(M).$$

Thus if $M \in \mathrm{Mod}\text{-}R$ is Γ_U-reflexive, then M is D-reflexive, and hence finitely generated. But then $\Gamma_U(M)$ is finitely generated, and M is U-torsion since

$$\Delta_{U_R} M \cong \Delta_{U_R} \Gamma_{sU} \Gamma_{U_R}(M) \cong HDDT' \Gamma_{U_R}(M) = 0. \qquad \blacksquare$$

Now we are in position to establish the connection between GMD's and cotilting modules over artin algebras.

Theorem 5.5.11. *Let R be an artin algebra and let U_R be a finitely generated module with $S = \mathrm{End}(U_R)$. Then U_R is a cotilting module if and only if $_S U_R$ induces a generalized Morita duality.*

Proof. If U_R is a finitely generated cotilting module, then according to Theorem 5.5.10(1), the U-reflexive modules are just the finitely generated U-torsionless modules. Thus $_S U_R$ induces a GMD.

The converse follows from Corollary 5.5.8, since a GMD is a WMD. $\qquad \blacksquare$

Actually, Morita duality and the cotilting dualities of Theorem 5.5.11 are the only examples of generalized Morita dualities that we are aware of.

5.6. Finitistic Cotilting Modules and Bimodules

We now consider WMD's in the noetherian case. We begin with a specialized definition of a cotilting module.

Definition 5.6.1. A *finitistic cotilting module* is a module U_R in mod-R such that

(1) inj . dim .$U_R \le 1$;
(3) $\mathrm{Ext}_R^1(U, U) = 0$;
(3) $\mathrm{Ker}(\Delta_{U_R}) \cap \mathrm{Ker}(\Gamma_{U_R}) \cap$ mod-$R = 0$.

We pointed out in Example 3.7.12 and Proposition 5.4.5 that noetherian serial rings provide examples of tilting modules and cotilting bimodules over non-artinian rings. They also supply examples of finitistic cotilting modules.

Proposition 5.6.2. *If R is right noetherian and right hereditary, and V_R is a tilting module, then V_R is a finitistic cotilting module.*

Proof. Suppose V_R is a tilting module that is not a finitistic cotilting module; then by Definition 5.6.1 there is a non-zero, finitely generated module M_R such that $\Delta_V(M) = 0 = \Gamma_V(M)$. Also by Theorem 3.2.1 there is an exact sequence

$$0 \to R \to V_0 \to V_1 \to 0$$

with $V_i \in \mathrm{add}(V_R)$. Hence $\mathrm{Hom}_R(M, R) = 0$, and the exactness of

$$\mathrm{Hom}_R(M, V_1) \to \mathrm{Ext}_R^1(M, R) \to \mathrm{Ext}_R^1(M, V_0)$$

shows that $\mathrm{Ext}_R^1(M, R) = 0$. But since R is right hereditary, considering a projective presentation of M_R, we see that this is not possible. ∎

If R is an artin algebra and U_R is a finitely generated cotilting module, then in the exact sequence $0 \to U_1 \longrightarrow U_0 \longrightarrow C \to 0$ of Theorem 5.2.5 one can have the $U_i \in \mathrm{add}(U_R)$. Thus a dual argument serves to verify the converse of Proposition 5.6.2 when R is a hereditary artin algebra.

Lemma 5.6.3. *If $_S U_R$ is a bimodule such that U_R and $_S U$ are noetherian, then $\mathrm{Cogen}(U_R) \cap$ mod-$R = \mathrm{cogen}(U_R)$ and $\mathrm{Cogen}(U_R) \cap S$-mod $= \mathrm{cogen}(_S U)$.*

Proof. The proof is straightforward. ∎

Regarding the finitely generated reflexive modules relative to a finitistic cotilting module we have

Proposition 5.6.4. *Suppose R is right noetherian and U_R is a finitistic cotilting module with $\text{End}(U_R) = S$ such that $_SU$ is noetherian. If M is a module in* mod-R, *then the following are equivalent.*

    (a) M_R is U-reflexive;
    (b) M_R is U-torsionless;
    (c) $\Gamma(M_R) = 0$.

Proof. $(a) \Rightarrow (b)$ is clear. Assuming (b) and referring to Lemma 5.6.3, the hypothesis that U_R is a finitistic cotilting module clearly implies (c). Similarly, assuming (c), we see that $\Gamma(\text{Rej}_U(M_R)) = 0$, and since (b) implies (c) we have $\Gamma(M/\text{Rej}_U(M_R)) = 0$. Hence the canonical sequence

$$0 \to \Delta(M/\text{Rej}_U(M_R)) \to \Delta(M) \to \Delta(\text{Rej}_U(M_R)) \to 0$$

is exact, so we can conclude that $\Delta(\text{Rej}_U(M_R)) = 0$. Since U_R is a finitistic cotilting module and R is right noetherian, we conclude that $\text{Rej}_U(M_R) = 0$, so M_R is U-torsionless (and we have proved $(c) \Rightarrow (b)$). We now apply Lemma 4.2.1, noting that we can take A finite since $\Delta(M_R)$ embeds in a finite direct sum of copies of $_SU$, to obtain an exact sequence

$$0 \to M_R \xrightarrow{f} U^n \to L_R \to 0$$

such that $\Delta(f)$ is an epimorphism and $\Gamma L \hookrightarrow \Gamma(U^n) = 0$. But then since $(c) \Rightarrow (b)$, L_R is U-torsionless, so M_R is U-reflexive by Lemma 4.2.3. ∎

Now we can establish the connection between weak Morita duality and finitistic cotilting modules.

Theorem 5.6.5. *The following are equivalent for any bimodule $_SU_R$.*

    (a) $_SU_R$ induces a WMD between the classes of U-torsionless modules in mod-R *and S-mod .*
    (b) (i) $_SU_R$ is faithfully balanced and R_R, $_SS$, U_R, and $_SU$ are noetherian,
       (ii) $\text{Ext}_R^1(U, U) = 0 = \text{Ext}_S^1(U, U)$,
       (iii) inj . dim .$U_R \leq 1$ *and* inj . dim .$_SU \leq 1$.

(c) (i) $_SU_R$ is faithfully balanced and R_R, $_SS$, U_R, and $_SU$ are noetherian,

  (ii) $_SU_R$ induces a duality between the classes of U-torsionless modules in mod-R and S-mod.

(d) (i) R is right noetherian, S is left noetherian, and U_R and $_SU$ are finitely generated and faithful,

  (ii) Every finitely generated U-torsionless module is U-reflexive.

(e) (i) $_SU_R$ is faithfully balanced, R is right noetherian, and S is left noetherian,

  (ii) $_SU$ and U_R are finitistic cotilting modules.

(f) (i) $S = \mathrm{End}(U_R)$ is left noetherian, R is right noetherian, and $_SU \in S$-mod,

  (ii) U_R is a finitistic cotilting module.

Proof. $(a) \Rightarrow (b)$. (i) is clear since the classes of finitely generated U-reflexive modules are closed under submodules and include R_R and $_SS$. (ii) and (iii) follow from Proposition 5.5.3.

$(b) \Rightarrow (c)$. By Lemma 5.6.3 it suffices to show that all modules in $\mathrm{cogen}(U)$ are U-reflexive. Suppose $M \in \mathrm{cogen}(U_R)$. By (ii) and (iii) it follows that $\Gamma(M) = 0$. Since $\Delta(K) \in \mathrm{cogen}(_SU)$, we also have $\Gamma \Delta K = 0$. Hence Δ^2 preserves the exactness of an exact sequence of the form $0 \to K \to R^n \to M \to 0$, and since M is U-torsionless and R^n is U-reflexive, we obtain that M is U-reflexive.

$(c) \Rightarrow (a)$. Referring to Lemma 5.6.3, since $\mathrm{cogen}(U)$ is closed under submodules, it suffices to show that $\mathrm{cogen}(U)$ is closed under extensions, and, for this, it suffices to show that $\mathrm{cogen}(U) = \mathrm{Ker}\,\Gamma \cap \mathrm{mod}\text{-}R$ (or S-mod). Let $M \in \mathrm{mod}\text{-}R$ and suppose $0 \to K \xrightarrow{f} R^n \to M \to 0$ is exact. If $\Gamma(M) = 0$, then, since K is U-reflexive, M is U-torsionless by Lemma 4.2.3. On the other hand, if M is U-torsionless, then Δf is epic by Lemma 4.2.4; thus, $\Gamma(M) = 0$.

$(c) \Rightarrow (d)$ is immediate and, since the hypotheses imply that R_R, U_R, $_SS$, and $_SU$ are finitely generated and U-torsionless, $(d) \Rightarrow (c)$ is straightforward.

Now, (e) follows immediately from (b) and (a) via Theorem 5.5.6, and it is clear from the definition that (e) implies (b).

Since (e) implies (f) is clear, it suffices to show that (f) implies (d). Assume (f). Since $\Gamma(R_R) = 0$, R_R is U-reflexive by Proposition 5.6.4, so (d)(i) is satisfied. Also by Proposition 5.6.4, all U-torsionless modules in mod-R are U-reflexive. Suppose $_SN$ is a U-torsionless module in S-mod and let

$$0 \to K \xrightarrow{j} Q \to N \to 0$$

be an exact sequence in S-mod with Q projective. Then $_S Q$ is U-reflexive and $\text{Im}(\Delta(j))$ is U-reflexive since it is a submodule of $\Delta(_S K) \in \text{cogen}(U_R)$. Thus $\Delta(j)$ is epic by Lemma 4.2.4, so K is U-reflexive by Lemma 4.2.3. Finally, since $\Gamma \Delta(K) = 0$ by Proposition 5.6.4, we see that Δ^2 preserves the exactness of the preceding sequence and obtain that N is U-reflexive. ∎

As a corollary to Theorem 5.6.5 we have the following result, which is due to J. P. Jans [51], whose work has provided inspiration for many of the results of this section.

Corollary 5.6.6. *Let R be a noetherian ring. Then the bimodule $_R R_R$ induces a duality between the R-torsionless modules in* mod-R *and R-mod if and only if* inj $.$ dim $._R R \leq 1$ *and* inj $.$ dim $. R_R \leq 1$.

According to Corollary 5.5.8, if U_R is a finitely generated module with $S = \text{End}(U_R)$ over an artin algebra R, then $_S U_R$ induces a weak Morita duality if and only if U_R is a cotilting module. Thus Theorem 5.6.5 also yields

Corollary 5.6.7. *A finitely generated module U_R is a finitistic cotilting module over an artin algebra R if and only if U_R is a cotilting module.*

Definition 5.6.8. A bimodule $_S U_R$ satisfying the conditions of Theorem 5.6.5 is called a *finitistic cotilting bimodule*.

From Theorem 5.5.6 and Theorem 5.6.5(a) we obtain

Corollary 5.6.9. *If $_S U_R$ is a finitistic cotilting bimodule, then $_S U_R$ induces a cotilting theorem between* mod-R *and S-mod.*

Of course $_{\mathbb{Z}} \mathbb{Z}_{\mathbb{Z}}$ is a finitistic cotilting bimodule. The following corollary yields further examples. (See Examples 3.7.7 and 3.7.12.)

Corollary 5.6.10. *If R is a hereditary noetherian serial ring with tilting bimodule $_S V_R$, then $_S V_R$ is a finitistic cotilting bimodule.*

Proof. According to Proposition 5.6.2, V_R is a finitistic cotilting module, and by Propositions B.1.6 and B.1.7, S is noetherian and $_S V$ is finitely generated. Thus Theorem 5.6.5(f) applies. ∎

5.7. *U*-torsionless Linear Compactness

Recall that a module M is linearly compact if, for every inverse system of epimorphisms $M \xrightarrow{p_\lambda} M_\lambda$, the inverse limit $\varprojlim p_\lambda$ is an epimorphism. Here we shall consider a more general notion (a version of which appeared in [42]) and its connection to U-reflexivity in regard to weak and generalized Morita duality and cotilting bimodules. Many of the proofs in this section have their roots in the papers [26] and [29].

Definition 5.7.1. Let $_S U_R$ be a bimodule and let Δ denote the U-dual. A U-torsionless module M is said to be *U-torsionless linearly compact* if, for every inverse system of maps $\{M \xrightarrow{p_\lambda} M_\lambda\}$ such that M_λ is U-torsionless and $\Delta(\text{Coker}(p_\lambda)) = 0$ for all λ, the inverse limit $\varprojlim p_\lambda$ has $\Delta(\text{Coker}(\varprojlim p_\lambda)) = 0$.

If $_S U_R$ is a bimodule, we continue to let Δ denote the U-dual, and we begin with a lemma that applies to costar modules.

Lemma 5.7.2. *Let $_S U_R$ be a bimodule. If M is U-torsionless linearly compact and ΔM is a direct limit of a directed system $\{N_\lambda \xrightarrow{i_\lambda} \Delta M\}$ of U-reflexive submodules N_λ, then ΔM is U-reflexive.*

Proof. Given M and $\{N_\lambda \xrightarrow{i_\lambda} \Delta M\}$ as in the hypothesis, let $p_\lambda = \Delta i_\lambda \circ \delta_M :$ $M \to \Delta N_\lambda$ to obtain an inverse system $\{M \xrightarrow{p_\lambda} \Delta N_\lambda\}$ in which each ΔN_λ is U-torsionless, and let $\Delta N_\lambda \xrightarrow{c_\lambda} C_\lambda = \text{Coker } p_\lambda$. From the commutative diagram

$$
\begin{array}{ccccc}
0 \to & N_\lambda & \xrightarrow{i_\lambda} & \Delta M & = \quad \Delta M \\
& \delta_{N_\lambda} \downarrow & & \delta_{\Delta M} \downarrow & \quad \| \\
& \Delta^2 N_\lambda & \xrightarrow{\Delta^2 i_\lambda} & \Delta^3 M & \xrightarrow{\Delta(\delta_M)} \Delta M
\end{array}
$$

with exact rows, we see, since δ_{N_λ} is an isomorphism and $\Delta(\delta_M) \circ \delta_{\Delta M} = 1_{\Delta M}$, that $\Delta(p_\lambda) = \Delta(\delta_M) \circ \Delta^2 i_\lambda$ is a monomorphism. Thus it follows from the exact sequence

$$
0 \to \Delta C_\lambda \longrightarrow \Delta^2 N_\lambda \xrightarrow{\Delta(p_\lambda)} \Delta M
$$

that $\Delta(\text{Coker } p_\lambda) = 0$, so by hypothesis, $\Delta(\text{Coker}(\varprojlim p_\lambda)) = 0$. Now since $\varinjlim i_\lambda$ is an isomorphism and

$$
\varprojlim p_\lambda = \varprojlim(\Delta(i_\lambda) \circ \delta_M) \cong \Delta(\varinjlim i_\lambda) \circ \delta_M,
$$

we have

$$\mathrm{Coker}(\varprojlim p_\lambda) \cong \mathrm{Coker}\,\delta_M,$$

and so applying Δ to

$$M \xrightarrow{\delta_M} \Delta^2 M \to \mathrm{Coker}\,\delta_M \to 0,$$

we see that the epimorphism $\Delta(\delta_M)$ is an isomorphism and so then is $\delta_{\Delta M}$. ∎

As an immediate consequence of Lemma 5.7.2 we have

Proposition 5.7.3. *If U_R is a costar module with $S = \mathrm{End}(U_R)$ and M_R is U-torsionless linearly compact, then ΔM is $_S U_R$-reflexive.*

Proof. Since U_R is a costar module, by definition every finitely generated U-torsionless left S-module is U-reflexive. Thus Lemma 5.7.2 applies. ∎

The class of linear compact modules in Mod-R is closed under submodules, epimorphic images, and extensions (see [81]). Very little is known about the closure properties of the class of U-torsionless linearly compact modules; however, employing the next lemma we shall see that it is closed under U-torsionless epimorphic images.

Lemma 5.7.4. *Let $_S U_R$ be a bimodule. If M, N, and L are in Mod-R, $f \in \mathrm{Hom}_R(M, N)$, and $g \in \mathrm{Hom}_R(N, L)$, then there is an exact sequence*

$$\mathrm{Coker}\,f \to \mathrm{Coker}(g \circ f) \to \mathrm{Coker}\,g \to 0.$$

Hence

(1) $\mathrm{Coker}\,g \in \mathrm{Ker}\,\Delta$ if $\mathrm{Coker}(g \circ f) \in \mathrm{Ker}\,\Delta$,
(2) $\mathrm{Coker}\,g \circ f \in \mathrm{Ker}\,\Delta$ if $\mathrm{Coker}\,f$, $\mathrm{Coker}\,g \in \mathrm{Ker}\,\Delta$.

Proof. Applying the Snake Lemma to the commutative diagram with exact rows

$$
\begin{array}{ccccccc}
 & M & \xrightarrow{f} & N & \to & \mathrm{Coker}\,f & \to 0 \\
 & g \circ f \downarrow & & g \downarrow & & \overline{g} \downarrow & \\
0 \to & g \circ f(M) & \hookrightarrow & L & \to & \mathrm{Coker}(g \circ f) & \to 0
\end{array}
$$

where $\overline{g}$ is induced by g, we obtain that $\mathrm{Coker}\,g$ is isomorphic to $\mathrm{Coker}\,\overline{g}$, and hence we have the exact sequence as asserted. ∎

Now we have the following closure property of the class of U-torsionless linearly compact modules in Mod-R. It seems not even to be known whether the class is closed under finite direct sums.

Proposition 5.7.5. *Let $_SU_R$ be a bimodule. If M is U-torsionless linearly compact, L is U-torsionless, and $p : M \to L$ with Coker $p \in \text{Ker } \Delta$, then L is U-torsionless linearly compact.*

Proof. Suppose that $\{q_\lambda : L \to L_\lambda\}$ is an inverse system with Coker $q_\lambda \in$ Ker Δ for all λ. By Lemma 5.7.4(2), Coker$(q_\lambda \circ p) \in$ Ker Δ for all λ, and so since M is U-torsionless linearly compact, Coker$(\varprojlim(q_\lambda \circ p)) \in$ Ker Δ, that is, Coker$((\varprojlim q_\lambda) \circ p) \in$ Ker Δ. Hence Coker$(\varprojlim q_\lambda) \in$ Ker Δ by Lemma 5.7.4(1). ∎

In order to examine the connection between U-torsionless linearly compactness and weak Morita duality, we need the next two lemmas.

Lemma 5.7.6. *If M_R is $_SU_R$-reflexive, every U-torsionless M' admitting an exact sequence*

$$M \longrightarrow M' \longrightarrow C \to 0$$

with $\Delta C = 0$ is U-reflexive, and $\Delta \text{Coker}(\Delta i) = 0$ for every embedding $_SL \overset{i}{\hookrightarrow} \Delta M$, then M is U-torsionless linearly compact.

Proof. Suppose that $\{M \overset{p_\lambda}{\longrightarrow} M_\lambda\}$ is an inverse system of maps with $M_\lambda \in$ Cogen(U_R), and $C_\lambda = \text{Coker } p_\lambda$ such that each $\Delta C_\lambda = 0$. From the latter it follows that each map in the direct system $\{\Delta M_\lambda \overset{\Delta p_\lambda}{\longrightarrow} \Delta M\}$ is monic. Thus we have an exact sequence

$$0 \to \varinjlim \Delta M_\lambda \overset{\varinjlim \Delta p_\lambda}{\longrightarrow} \Delta M.$$

So by hypothesis $\Delta(\text{Coker}(\Delta(\varinjlim \Delta p_\lambda))) = 0$. In the commutative diagram

$$
\begin{array}{ccccc}
\Delta^2 M & \overset{\Delta(\varinjlim \Delta p_\lambda)}{\longrightarrow} & \Delta(\varinjlim \Delta M_\lambda) & \longrightarrow & \text{Coker}(\Delta(\varinjlim \Delta p_\lambda)) \to 0 \\
\delta_M \uparrow & & \uparrow & & \\
M & \overset{\varprojlim p_\lambda}{\longrightarrow} & \varprojlim M_\lambda & \longrightarrow & \text{Coker}(\varprojlim p_\lambda) \quad \to 0
\end{array}
$$

with exact rows, the vertical maps are both isomorphisms since M and the M_λ are U-reflexive and $\Delta(\varinjlim \Delta M_\lambda) \cong \varprojlim(\Delta^2 M_\lambda)$. But then we can complete the diagram to obtain an isomorphism to which, applying Δ, we have

$$\Delta(\text{Coker}(\varprojlim p_\lambda)) \cong \Delta(\text{Coker}(\Delta(\varprojlim \Delta p_\lambda))) = 0. \qquad \blacksquare$$

Theorem 5.7.7. *If $_SU_R$ induces a WMD between $\mathcal{F}_R$ and $_S\mathcal{F}$, then every module in $\mathcal{F}_R$ and $_S\mathcal{F}$ is U-torsionless linearly compact.*

Proof. Suppose that $M \in \mathcal{F}_R$, and $0 \to M \xrightarrow{p} M' \longrightarrow C \to 0$ with M' U-torsionless and $\Delta C = 0$. From the latter it follows that $\Delta M' \xrightarrow{\Delta p} \Delta M$ is monic, so $\Delta M'$ is U-reflexive. Thus by Lemma 5.3.5, Proposition 5.5.4, and Proposition 5.5.3, M' is U-reflexive. If we have an exact sequence

$$0 \to L \xrightarrow{i} \Delta M \to X \to 0,$$

then, since by Proposition 5.5.4 $\Gamma \Delta M = 0$, we see that $\text{Coker}(\Delta i) \cong \Gamma X$. But also by Proposition 5.5.4, $\Delta \Gamma X = 0$. Thus, Lemma 5.7.6 applies. $\blacksquare$

Now we have the tools to characterize WMD's.

Theorem 5.7.8. *Let $\mathcal{F}_R$ and $_S\mathcal{F}$ be subcategories of Mod-R and S-Mod that are closed under submodules and extensions and contain R_R and $_SS$, respectively. Then the $_SU_R$-dual Δ induces a WMD between $\mathcal{F}_R$ and $_S\mathcal{F}$ if and only if the following five conditions hold:*

(1) $_SU_R$ is faithfully balanced;
(2) $\Delta : \mathcal{F}_R \rightleftarrows {}_S\mathcal{F} : \Delta$;
(3) $\Gamma M = 0$, if $M \in \mathcal{F}_R$, and $\Gamma N = 0$, if $N \in {}_S\mathcal{F}$;
(4) $\text{Ker}\,\Delta \cap \text{Ker}\,\Gamma \cap \text{gen}(\mathcal{F}_R) = 0$ and $\text{Ker}\,\Delta \cap \text{Ker}\,\Gamma \cap \text{gen}(_S\mathcal{F}) = 0$;
(5) All modules in $\mathcal{F}_R$ and $_S\mathcal{F}$ are U-torsionless linearly compact.

Proof. We know that all the conditions are necessary by Propositions 4.1.5, 5.5.3, and 5.5.4 and Theorem 5.7.7.

For sufficiency, let us denote both $\mathcal{F}_R$ and $_S\mathcal{F}$ by $\mathcal{F}$.

Suppose that $X \in \text{gen}(\mathcal{F})$ so that there is an exact sequence

$$0 \to K \xrightarrow{f} M \to X \to 0$$

with M and K in $\mathcal{F}$. Then, from the exact sequence $0 \to \Delta X \to \Delta M$ we have $\Delta X \in \mathcal{F}$, $\Delta^2 M \in \mathcal{F}$ and, since it embeds in $\Delta^2 X$, $X/\text{Rej}_U(X) \in \mathcal{F}$. In particular, if $X \in \text{Cogen}(U)$, then $X \in \mathcal{F} \subseteq \text{Ker}\,\Gamma$. Now from the exact sequence

$0 \to \mathrm{Rej}_U(X) \longrightarrow X \longrightarrow X/\mathrm{Rej}_U(X) \to 0$ we obtain an exact sequence

$$\Delta X \to \Delta(\mathrm{Rej}_U(X)) \to \Gamma(X/\mathrm{Rej}_U(X)) = 0$$

from which it follows that $\Delta(\mathrm{Rej}_U(X)) = 0$. Since right ideals of R (or left ideals of S) belong to $\mathcal{F}$, it follows, as in the proof of Proposition 5.5.3(2) that $\mathrm{Id}(U) \leq 1$. Thus if $\Gamma X = 0$, we have an exact sequence

$$0 = \Gamma X \to \Gamma(\mathrm{Rej}_U(X)) \to \mathrm{Ext}^2(X/\mathrm{Rej}_U(X)) = 0,$$

and so we have shown that $X \in \mathrm{Cogen}(U)$ if and only if $\Gamma X = 0$. Now from the exact sequence

$$\Delta M \xrightarrow{\Delta f} \Delta K \longrightarrow \Gamma X \to 0$$

we see that Δf is epic if and only $X \in \mathrm{Cogen}(U)$. In particular, letting $M = U^n$, it follows from Theorem 4.2.7(e) that U is a costar module, and thus every finitely generated U-torsionless module is U-reflexive. In particular, if $M \in \mathcal{F}$, then every finitely generated submodule of ΔM is U-reflexive. Thus by Lemmas 5.7.2 and 5.3.5 and condition (4) every module in $\mathcal{F}$ is U-reflexive, and $_S U_R$ induces a WMD between $\mathcal{F}_R$ and $_S\mathcal{F}$. ∎

Lemma 5.7.9. *Let $_S U_R$ be a bimodule. Assume that U-torsionless modules in* mod-R *are U-reflexive and suppose $M \in$ Mod-R is U-torsionless and ΔM is U-torsionless linearly compact. If $L \xhookrightarrow{i} M$, then $\Delta(\mathrm{Coker}\,\Delta i) = 0$, and so ΔL is U-torsionless linearly compact.*

Proof. Let $\{L_\lambda \xhookrightarrow{i_\lambda} M\}$ be the upward directed family of finitely generated submodules of L so that $L = \varinjlim L_\lambda$. Then we obtain exact sequences

$$\Delta M \xrightarrow{\Delta i_\lambda} \Delta L_\lambda \to \mathrm{Coker}\,\Delta(i_\lambda) \to 0,$$

and since M is U-torsionless and the L_λ are U-reflexive by hypothesis, it follows that $\Delta(\mathrm{Coker}\,\Delta(i_\lambda)) = 0$. Since ΔM is U-torsionless linearly compact, we obtain $\Delta(\mathrm{Coker}\,\varprojlim\Delta(i_\lambda)) = 0$. But

$$\Delta(\mathrm{Coker}\,\varprojlim\Delta(i_\lambda)) \cong \Delta(\mathrm{Coker}\,\Delta(\varinjlim i_\lambda)) \cong \Delta(\mathrm{Coker}(\Delta i)).$$

And so, by Proposition 5.7.5, ΔL is U-torsionless linearly compact. ∎

If $_S U_R$ induces a Morita duality, then U_R and $_S U$ are injective cogenerators (Theorem 4.4.2). Thus, in this case, U-torsionless linearly compact is simply linearly compact, $\Gamma M = 0$ for all M, and $\mathrm{Ker}\,\Delta = 0$.

Our concluding characterization of generalized Morita duality complements the characterization, Corollary 4.4.4, of the reflexive modules under a Morita duality as the linearly compact modules.

Theorem 5.7.10. *Let $_SU_R$ be a bimodule. Let $\mathcal{F}_R$ and $_S\mathcal{F}$ denote the classes of all U-reflexive modules in* Mod-R *and* S-Mod. *Then U induces a GMD if and only if*

(1) $_SU_R$ is faithfully balanced,

(2) $\Gamma(M) = 0$ for all M in $\mathcal{F}_R$ and $\Gamma(N) = 0$ for all N in $_S\mathcal{F}$,

(3) $\mathrm{Ker}\,\Delta \cap \mathrm{Ker}\,\Gamma \cap \mathrm{gen}(\mathcal{F}_R) = 0$ *and* $\mathrm{Ker}\,\Delta \cap \mathrm{Ker}\,\Gamma \cap \mathrm{gen}(_S\mathcal{F}) = 0$,

(4) $\mathcal{F}_R$ and $_S\mathcal{F}$ consist precisely of the U-torsionless linearly compact modules in Mod-R *and* S-Mod.

Proof. First assume that U induces a GMD. Since R_R and $_SS$ are U-reflexive, $_SU_R$ is faithfully balanced. Then (2) and (3) follow from Theorem 5.7.8. Also by Theorem 5.7.8, for (4), we need only show that any U-torsionless linearly compact module is U-reflexive. According to Proposition 5.5.7, U_R and $_SU$ are costar modules. Hence if M in Mod-R is U-torsionless linearly compact, then ΔM is U-reflexive by Proposition 5.7.3. But then M is U-reflexive by Lemma 5.3.5 and Proposition 5.5.3.

Conversely, assume (1)–(4). By (1), R_R and $_SS$, are U-reflexive. By the symmetry of the hypotheses it suffices to prove that $\mathcal{F}_R$ is closed under submodules and extensions. For closure under extensions suppose that

$$0 \to M_1 \to M \to M_2 \to 0$$

is exact in Mod–R where M_1 and M_2 are U-reflexive. Since $\Gamma M_2 = 0$ and $\Gamma \Delta M_1 = 0$ by (2), we can apply Δ twice to obtain the straightforward proof that M is U-reflexive. Finally, suppose that M in Mod-R is U-reflexive and that L is a submodule of M. Applying Δ to $L \overset{i}{\hookrightarrow} M$ we have the exact sequence

$$\Delta M \overset{\Delta i}{\to} \Delta L \to \mathrm{Coker}\,\Delta i \to 0.$$

Note that, since finitely generated free modules are U-reflexive and hence U-torsionless linearly compact by (4), we can conclude from Lemma 5.7.5 that finitely generated U-torsionless modules are U-torsionless linearly compact and therefore U-reflexive, again by (4). Since ΔM is U-reflexive, it is U-torsionless linearly compact by (4). Hence $\Delta(\mathrm{Coker}\,\Delta i) = 0$ by Lemma 5.7.9, and ΔL is U-torsionless linearly compact by Lemma 5.7.5. Thus

ΔL is U-reflexive by (4), and so L is U-reflexive by Lemma 5.3.5 and condition (3). ∎

Several of the preceding Lemmas also prove useful in characterizing the U-torsionless linearly compact modules when U_R is a cotilting module.

Theorem 5.7.11. *Let U_R be a cotilting module with $S = \text{End}(U_R)$. Then M_R is U-torsionless linearly compact if and only if M_R is $_SU_R$-reflexive and $\text{Coker}\,\Delta i \in \text{Ker}\,\Delta$ whenever $L \overset{i}{\hookrightarrow} \Delta M$ in S-Mod.*

Proof. ($\Rightarrow$) Suppose that M_R is U-torsionless linearly compact. Since the cotilting module U_R is a costar module by Proposition 5.2.3, satisfies $\Gamma\Delta^2 = 0$ by definition, and has $\text{Ker}\,\Delta \cap \text{Ker}\,\Gamma = 0$ by Proposition 5.2.6, the fact that M is reflexive follows from Proposition 5.7.3 and Lemma 5.3.5. That $\text{Coker}\,\Delta i \in \text{Ker}\,\Delta$ whenever $L \overset{i}{\hookrightarrow} \Delta M$ follows from Lemma 5.7.9 (with R and S interchanged) since M is U-reflexive.

($\Leftarrow$) Assuming the conditions, according to Lemma 5.7.6 we need only show that a U-torsionless module M'_R is $_SU_R$-reflexive whenever there is an exact sequence

$$M \xrightarrow{p} M' \xrightarrow{g} C \to 0$$

in which M is $_SU_R$-reflexive and $\Delta C = 0$. To this end, let $K = \text{Ker}\,p$ and $I = \text{Im}\,p$ and consider the exact sequences

$$0 \to K \xrightarrow{f} M \xrightarrow{\alpha} I \to 0$$

and

$$0 \to I \xrightarrow{\beta} M' \xrightarrow{g} C \to 0$$

with $\beta \circ \alpha = p$. Since $I \in \text{Cogen}(U_R) = \text{Ker}\,\Gamma$, we see that Δf is epic. Thus K is U-reflexive by Lemma 4.2.3. From the commutative diagram

$$
\begin{array}{ccccc}
M & \xrightarrow{\alpha} & I & \longrightarrow & 0 \\
\delta_M \downarrow & & \delta_I \downarrow & & \\
\Delta^2 M & \xrightarrow{\Delta^2\alpha} & \Delta^2 I & \longrightarrow & \text{Coker}(\Delta^2\alpha) \to 0
\end{array}
$$

we see that $\text{Coker}(\Delta^2\alpha) \cong \text{Coker}\,\delta_I$. By hypothesis $\Delta(\text{Coker}(\Delta^2\alpha)) = 0$ since $\Delta I \overset{i}{\hookrightarrow} \Delta M$, and $\Gamma(\text{Coker}\,\delta_I) = 0$ by Lemma 5.3.5. Thus by Proposition 5.2.6 $\text{Coker}\,\delta_I = 0$, so I is U-reflexive. Since $\Delta C = 0$ and $\Gamma M' = 0$, we obtain a commutative diagram with exact rows

$$
\begin{array}{ccccccc}
0 \to & I & \stackrel{\beta}{\longrightarrow} & M' & \to C \to 0 \\
& \delta_I \downarrow \cong & & \delta_{M'} \downarrow & \\
& \Delta^2 I & \stackrel{\Delta^2\beta}{\longrightarrow} & \Delta^2 M' & \to C' \to 0
\end{array}
$$

in which $C' = \mathrm{Coker}(\Delta^2\beta)$. Since $\Delta C = 0$ we have $\Delta M' \stackrel{\Delta p}{\longleftrightarrow} \Delta M$, so by hypothesis $\Delta \mathrm{Coker}(\Delta^2 p) = 0$. Thus, since $\Delta^2 p = \Delta^2\beta \circ \Delta^2\alpha$ we see from Lemma 5.7.4 that $\Delta C' = 0$. Now we can apply Δ to obtain the commutative diagram with exact rows

$$
\begin{array}{ccccc}
0 \to & \Delta^3 M' & \stackrel{\Delta^3\beta}{\longrightarrow} & \Delta^3 I \\
& \Delta\delta_{M'} \downarrow & & \Delta\delta_I \downarrow \cong \\
0 \to & \Delta M' & \stackrel{\Delta^2\beta}{\longrightarrow} & \Delta I
\end{array}
$$

from which it follows that the epimorphism $\Delta\delta_{M'}$ is actually an isomorphism, and then so is $\delta_{\Delta M'}$, since $\Delta\delta_{M'} \circ \delta_{\Delta M'} = 1_{\Delta M'}$. Finally, since $\Delta M'$ is $_S U_R$-reflexive and $\mathrm{Ker}\,\Delta \cap \mathrm{Ker}\,\Gamma = 0$, so is M' by Lemma 5.3.5. ∎

Since a cotilting bimodule satisfies the first three conditions of Theorem 5.7.10, these last two theorems easily yield

Corollary 5.7.12. *A cotilting bimodule $_S U_R$ induces a GMD if and only every U-reflexive module is U-torsionless linearly compact.*

5.8. Examples and Questions

In this section we shall present examples showing that, although, as we have seen, they all induce cotilting theorems, cotilting bimodules, bimodules inducing weak Morita dualities, and those inducing generalized Morita dualities, are all distinct notions.

It follows from Lemma 5.6.3 and Theorem 5.6.5 that if R is a hereditary noetherian ring, then $_R R_R$ induces a weak Morita duality between the projective modules in mod-R and R-mod. The ring $\mathbb{Z}$ of integers is such a ring, and $_\mathbb{Z}\mathbb{Z}_\mathbb{Z}$ is neither a cotilting bimodule nor does it induce a generalized Morita duality.

Example 5.8.1. *According to Theorem 5.6.5, $_\mathbb{Z}\mathbb{Z}_\mathbb{Z}$ induces a (self) WMD between the finitely generated $\mathbb{Z}$-torsionless (i.e., the finitely generated free modules). However $\Gamma(\mathbb{Z}^\mathbb{N}) = \mathrm{Ext}^1_\mathbb{Z}(\mathbb{Z}^\mathbb{N}, \mathbb{Z}) \neq 0$ (i.e., $Z^\mathbb{N}$ is not a Whitehead*

group [36, Theorem 95.3, page 165 and Proposition 99.2, page 179]), so $_{\mathbb{Z}}\mathbb{Z}_{\mathbb{Z}}$
is not a cotilting bimodule. On the other hand, $_{\mathbb{Z}}\mathbb{Z}_{\mathbb{Z}}$ *does not induce a GMD
by Theorem 5.7.10, since* $\mathbb{Z}^{\mathbb{N}} = \Delta(\mathbb{Z}^{(\mathbb{N})})$ *is* $\mathbb{Z}$*-reflexive ([35, Corollary 2.5,
page 61]), but* $\Gamma(\mathbb{Z}^{\mathbb{N}}) \neq 0$.

There are non-noetherian examples of domains that induce weak
Morita dualities. Indeed, according to a theorem of E. Matlis, (see [37,
Proposition 5.6, page 145]), if D is an integral domain with quotient field
Q, then every module in cogen(D) is D reflexive if and only if Q/D is an
injective cogenerator in Mod-D. In view of the following proposition we
see that in this case $_DD_D$ induces a WMD on cogen(D). (According to [37,
Proposition 5.7, page 146], there are non-noetherian domains satisfying this
condition.)

Proposition 5.8.2. *Let W be a module over a ring R. If* $\mathrm{Ext}_R^1(W, W) = 0$
and inj . dim .$(W) \leq 1$, *then* cogen(W) *is contained in* $\mathrm{Ker}\,\mathrm{Ext}_R^1(\_, W)$ *and is
closed under extensions.*

Proof. Suppose first that there is an exact sequence

$$0 \to M \longrightarrow W^n \longrightarrow C \to 0.$$

This yields an exact sequence

$$\mathrm{Ext}_R^1(W^n, W) \to \mathrm{Ext}_R^1(M, W) \to \mathrm{Ext}_R^2(C, W),$$

which, by hypothesis, shows that $\mathrm{Ext}_R^1(M, W) = 0$.

Let

$$0 \to M_1 \xrightarrow{f} X \xrightarrow{g} M_2 \to 0$$

be exact and suppose that also $M_1 \hookrightarrow W^m$ and $M_2 \hookrightarrow W^n$ so that we
have maps $\theta_1, \ldots, \theta_n \in \mathrm{Hom}_R(M_2, W)$ and $\varphi_1, \ldots, \varphi_m \in \mathrm{Hom}_R(M_1, W)$
with $\cap_1^n \mathrm{Ker}\,\theta_i = 0$ and $\cap_1^m \mathrm{Ker}\,\varphi_i = 0$. Since $\mathrm{Ext}_R^1(M_2, W) = 0$, we see that
$\varphi_1, \ldots, \varphi_m$ extend to $\overline{\varphi}_1, \ldots, \overline{\varphi}_m \in \mathrm{Hom}_R(X, W)$. Thus, if $x \in (\cap_1^n \mathrm{Ker}\,\theta_i \circ
g) \cap (\cap_1^m \mathrm{Ker}\,\overline{\varphi}_i)$ we see that $x = 0$ so that $X \hookrightarrow W^{m+n}$. ∎

We began this section by recalling that hereditary noetherian rings induce
weak Morita dualities. Next we shall show that, if R is a hereditary perfect
ring, then $_RR_R$ is a cotilting bimodule for which the reflexive modules are the
finitely generated projective modules, and thus over such a ring the reflexive
modules need not be closed under submodules. We begin with

Proposition 5.8.3. *If R is a right perfect right hereditary left coherent ring, then R_R is a cotilting module and $\mathrm{Ext}^1_R(M, R_R) = 0$ if and only if M is projective.*

Proof. Since R is right hereditary, every right R-module has injective dimension ≤ 1. Since direct products of projective right R-modules are projective (see [1, Theorems 19.20 and 28.4]), $\mathrm{Ext}^1_R(R^A_R, R_R) = 0$ for any set A. If $M \in \mathrm{Mod}\text{-}R$, then the terms P_1 and P_0 in a projective resolution

$$0 \to P_1 \longrightarrow P_0 \longrightarrow M \to 0$$

of M belong to $\mathrm{Prod}(R_R)$ by Lemma 1.3.1. Thus, letting M be a cogenerator, we see that the conditions of Theorem 5.2.5 are verified and R_R is a cotilting module. On the other hand, if $\mathrm{Ext}^1_R(M, R_R) = 0$, then since $\mathrm{Ext}^1_R(M, \_)$ commutes with direct products ([69, Theorem 7.14]), the sequence splits. Thus $\mathrm{Ext}^1_R(M, R_R) = 0$ if and only if M is projective. ∎

Our next proposition may be of independent interest, but it is nearly all the more we need to obtain the desired example.

Proposition 5.8.4. *If R is a basic two-sided perfect coherent ring, then a projective R-module P is $_R R_R$-reflexive if and only if P is finitely generated.*

Proof. Finitely generated projective R modules are reflexive over any ring R ([1, Proposition 20.17]). So assume that R is a two-sided perfect coherent ring with radical J and that P_R is projective but not finitely generated. According to [1, Propositions 20.13 and 28.13] we may assume that $P = eR^{(A)}$ for some primitive idempotent e in a basic set of primitive idempotents $\{e_1, \dots e_n\}$ for R and some infinite set A. Then $\Delta P \cong Re^A$. Since $J(Re^A) \subseteq Je^A$, we see that $(Re/Je)^A$ is semisimple, and since e is the only member of $\{e_1, \dots, e_n\}$ that does not annihilate it, $(Re/Je)^A$ is a vector space over the division ring $R/\ell_R(Re/Je)$, which is isomorphic to Re/Je as left R-modules. But then by [50, Theorem 2, Page 247], $(Re/Je)^A \cong (Re/Je)^{(B)}$ with $\mathrm{card}(B) > \mathrm{card}(A)$. Thus there is an epimorphism $Re^A \to (Re/Je)^{(B)}$, so, since Re^A is projective, the projective cover $Re^{(B)}$ of $(Re/Je)^{(B)}$ ([1, Proposition 28.13]) is a direct summand of $\Delta P \cong Re^A$ ([1, Proposition 17.15]). But then, a similar argument shows that the projective module $\Delta^2 P$ has a direct summand isomorphic to $eR^{(C)}$ with $\mathrm{card}(C) > \mathrm{card}(B) > \mathrm{card}(A)$, and thus it cannot be isomorphic to $P = eR^{(A)}$ according to [1, Theorem 28.14.]. ∎

Immediately now, in view of Theorem 5.3.6, we have

Corollary 5.8.5. *If R is a two-sided perfect hereditary ring, then $_R R_R$ is a cotilting bimodule that induces a cotilting theorem between the categories of finitely presented modules $\mathcal{A}_R$ and $_R\mathcal{A}$ with torsion theories $(\mathcal{T}_R, \mathcal{F}_R)$ and $(_R\mathcal{T}, _R\mathcal{F})$ in which classes of the R-reflexive modules $\mathcal{F}_R$ and $_R\mathcal{F}$ are just the projective modules in* mod-R *and* R-mod.

Now we can easily verify an example, due to G. D'Este [33], of a finitely generated cotilting bimodule that does not induce a weak Morita duality.

Example 5.8.6. *Let L be an infinite dimensional vector space over a field K, and let*

$$R = \begin{bmatrix} K & 0 \\ L & K \end{bmatrix},$$

the ring of lower triangular matrices with entries as indicated. Then $_R R_R$ is a cotilting bimodule whose reflexive modules are precisely the R-torsionless modules in mod-R *and* R-mod, *but $_R R_R$ does not induce a WMD.*

Of course, we already know from Corollary 5.7.12 that the algebra R of Example 5.8.6 must have an R-reflexive module that is not R-torsionless linearly compact. In [33, Corollary 2.6] D'Este characterized the R-torsionless linearly compact R-reflexive modules as those finitely generated projective modules that contain no finite dimensional indecomposable direct summands. Thus, though they are both R-reflexive, if e_1, e_2 denote the primitive diagonal idempotents in R, then $e_2 R$ is R-torsionless linearly compact, while the simple projective right R-module $e_1 R$ is not.

Although we have made use of the notion of U-torsionless linearly compactness in the preceding section, we view it as a rather mysterious one. For example, we would be interested to see a direct proof that $\mathbb{Z}$ is $\mathbb{Z}$-linearly compact.

The fact that the U-reflexive modules relative to a Morita duality bimodule $_S U_R$ are just the linearly compact modules suggests the problem of determining the classes of U-reflexive modules when $_S U_R$ merely induces a weak Morita duality. Are they the U-torsionless linearly compact modules? In particular, since $\mathbb{Z}^{(\mathbb{N})}$ and $\mathbb{Z}^{\mathbb{N}}$ are $\mathbb{Z}$-reflexive, we wonder if they are $\mathbb{Z}$-torsionless linearly compact.

As we have just suggested, given a bimodule $_S U_R$, the class of U-torsionless linearly compact R-modules is not well understood. That it is closed under U-torsionless factor modules is a consequence of Lemma 5.7.5. The known properties of the class of linearly compact modules (see [81, Section 3]) lead us to ask:

Is the class of U-torsionless linearly compact R-modules closed under

(a) submodules?
(b) U-torsionless extensions?
(c) finite direct sums?
(d) any infinite direct sums?

Proposition 5.8.7. *Let* $_SU_R$ *be a cotilting bimodule. If S is left noetherian, then* U_R *is* U-*torsionless linearly compact.*

Proof. Since U_R is reflexive and $_SS \cong \Delta(U_R)$, according to Theorem 5.7.11 it will suffice to observe that Coker $\Delta i \in$ Ker Δ for every left ideal I of S and inclusion map $i : I \to {}_SS$. But, since $\Gamma(_SS) = 0$, there is an exact sequence $\Delta(_SS) \xrightarrow{\Delta i} \Delta I \to \Gamma(S/I) \to 0$, and since I, being finitely generated and U-torsionless, and $_SS$ are reflexive, $S/I \in {}_S\mathcal{A}$ in Theorem 5.3.6. Thus (see Definition 5.1.1) Coker $\Delta i \cong \Gamma(S/I) \in$ Ker Δ. $\blacksquare$

In view of this proposition and of Corollary 5.7.12 and Theorem 5.7.8, we wonder (as was suggested to us by R. Colpi) if the cotilting bimodules of Proposition 5.4.5 (or more specially the one of Example 5.4.6) might induce a GMD or a WMD. In particular, are the modules R_R and $_SS$ U-torsionless linearly compact in these cases?

Results in Sections 4.3 and 4.4 suggest the presently open problems of determining one-sided characterizations of weak and generalized Morita duality and cotilting bimodules.

Appendix A
Adjoints and Category Equivalence

In this appendix we will give an exposition of some results from category theory that suffice to justify our assumption in Chapter 2 that the functors of interest can be taken to be Hom and tensor functors.

Let C be a category. If X is an object of C we write $X \in Ob\ C$ or just $X \in C$ if no confusion results. If $X, Y \in C$ we denote the set of morphisms with domain X and codomain Y by $\mathrm{Hom}_C(X, Y)$. We write elements in $\mathrm{Hom}_C(X, Y)$ in the form $X \xrightarrow{f} Y$ or $f : X \to Y$ and we denote the identity morphism in $\mathrm{Hom}_C(X, X)$ by 1_X. If $\mathcal{D}$ is another category and F is a functor from C to $\mathcal{D}$ we write $F : C \to \mathcal{D}$; and $F : C \rightleftarrows \mathcal{D} : G$ is shorthand for $F : C \to \mathcal{D}$ and $G : \mathcal{D} \to C$. We denote the identity functor on C by 1_C. If X is an object of C, then $\mathrm{Hom}_C(X, \_)$ and $\mathrm{Hom}_C(\_, X)$ denote the canonically induced (covariant and contravariant) functors from C into the category Set of all sets.

A.1. The Yoneda–Grothendieck Lemma

Lemma A.1.1 (Yoneda Lemma). *Suppose X is an object in the category C and $F : C \to Set$ is a covariant functor. If $u \in FX$, letting*

$$\Theta(u)_Z : \mathrm{Hom}_C(X, Z) \to FZ \ \text{ for each } \ Z \in C$$

via

$$\Theta(u)_Z : f \longmapsto Ff(u) \ \text{ for } \ f \in \mathrm{Hom}_C(X, Z),$$

one obtains a natural transformation $\Theta(u)$ from $\mathrm{Hom}_C(X, \_)$ to F. The mapping Θ defines a bijection between FX and the set of all natural transformations from $\mathrm{Hom}_C(X, \_)$ to F that is natural in both X and F. Moreover, if τ is such a natural transformation, then $\tau_X(1_X) \mapsto \tau$ under this correspondence.

Proof. It is straightforward that $\Theta(u)$, as defined, is a natural transformation. Let $\Phi(\tau) = \tau_X(1_X)$ if τ is a natural transformation. Then

$$\Phi\Theta(u) = (\Theta(u))_X(1_X) = F(1_X)(u) = 1_{FX}(u) = u,$$

so $\Phi\Theta = 1_{FX}$. Moreover,

$$\Theta\Phi(\tau) = \Theta(\tau_X(1_X)) = \tau$$

where the last equality follows from the commutativity of the diagram

$$\begin{array}{ccc} \mathrm{Hom}_{\mathcal{C}}(X, X) & \xrightarrow{\tau_X} & FX \\ \mathrm{Hom}_{\mathcal{C}}(X, f) \downarrow & & \downarrow Ff \\ \mathrm{Hom}_{\mathcal{C}}(X, Z) & \xrightarrow{\tau_Z} & FZ, \end{array}$$

that is, $(\Theta(\tau_X(1_X)))_Z(f) = Ff(\tau_X(1_X)) = \tau_Z(f)$, completing the proof that Θ and Φ are inverse functions. The proof that Θ is natural in X and F is straightforward. ∎

Corollary A.1.2. *If* $X, Y \in \mathcal{C}$, *then the correspondence* $u \mapsto \mathrm{Hom}_{\mathcal{C}}(u, \_)$, *where* $\mathrm{Hom}_{\mathcal{C}}(u, Z)(f) = fu$ *for* $u \in \mathrm{Hom}_{\mathcal{C}}(Y, X)$ *and* $f \in \mathrm{Hom}_{\mathcal{C}}(X, Z)$, *defines a bijection between* $\mathrm{Hom}_{\mathcal{C}}(Y, X)$ *and the set of natural transformations from* $\mathrm{Hom}_{\mathcal{C}}(X, \_)$ *to* $\mathrm{Hom}_{\mathcal{C}}(Y, \_)$ *that is natural in* X *and* Y. *Moreover, under this correspondence, isomorphisms correspond to natural isomorphisms.*

Proof. It suffices to note that u is an isomorphism if and only if $\mathrm{Hom}_{\mathcal{C}}(u, \_)$ is a natural isomorphism. If $\mathrm{Hom}_{\mathcal{C}}(u, \_)$ is an isomorphism, then $\mathrm{Hom}_{\mathcal{C}}(u, Y)$ is an isomorphism, so there exists $v \in \mathrm{Hom}_{\mathcal{C}}(X, Y)$ such that $vu = 1_Y$. Since $\mathrm{Hom}_{\mathcal{C}}(u, X)$ is an isomorphism, $\mathrm{Hom}_{\mathcal{C}}(u, X)(1_X) = u$, and $\mathrm{Hom}(u, X)(uv) = u$, we have $uv = 1_X$. Hence u is an isomorphism. ∎

A.2. Adjoint Covariant Functors

Adjoints and Arrows of Adjunction

Given categories $\mathcal{C}$ and $\mathcal{D}$ and a pair of covariant functors $H : \mathcal{C} \rightleftarrows \mathcal{D} : T$, T is a *left adjoint* of H and H is a *right adjoint* of T if, for each pair of objects N of $\mathcal{D}$ and M of $\mathcal{C}$, there is a bijection $\alpha_{N,M} : \mathrm{Hom}_{\mathcal{D}}(N, HM) \to \mathrm{Hom}_{\mathcal{C}}(TN, M)$ such that the diagrams

$$\begin{array}{ccc} \mathrm{Hom}_{\mathcal{D}}(N', HM') & \xrightarrow{\alpha_{N',M'}} & \mathrm{Hom}_{\mathcal{C}}(TN', M') \\ \mathrm{Hom}_{\mathcal{D}}(f, HM') \downarrow & & \downarrow \mathrm{Hom}_{\mathcal{C}}(Tf, M') \\ \mathrm{Hom}_{\mathcal{D}}(N, HM') & \xrightarrow{\alpha_{N,M'}} & \mathrm{Hom}_{\mathcal{C}}(TN, M') \end{array}$$

and

$$\begin{array}{ccc} \mathrm{Hom}_{\mathcal{D}}(N, HM) & \xrightarrow{\alpha_{N,M}} & \mathrm{Hom}_{\mathcal{C}}(TN, M) \\ \mathrm{Hom}_{\mathcal{D}}(N, Hg) \downarrow & & \downarrow \mathrm{Hom}_{\mathcal{C}}(TN, g) \\ \mathrm{Hom}_{\mathcal{D}}(N, HM') & \xrightarrow{\alpha_{N,M'}} & \mathrm{Hom}_{\mathcal{C}}(TN, M') \end{array}$$

are commutative for all morphisms $N \xrightarrow{f} N'$ in $\mathcal{D}$ and $M \xrightarrow{g} M'$ in $\mathcal{C}$. That is, if $N' \xrightarrow{v} HM'$ and $N \xrightarrow{u} HM$ are morphisms we have the equations

$$\alpha_{N',M'}(v) \circ Tf = \alpha_{N,M'}(v \circ f) \tag{A.1}$$

and

$$g \circ \alpha_{N,M}(u) = \alpha_{N,M'}(Hg \circ u). \tag{A.2}$$

In particular, for $N = HM$ and $u = 1_{HM}$, we have from (A.2)

$$g \circ \alpha_{HM,M}(1_{HM}) = \alpha_{HM,M'}(Hg)$$

and for $N' = HM'$, $f = Hg$, and $v = 1_{HM'}$, (A.1) becomes

$$\alpha_{HM,M'}(Hg) = \alpha_{HM',M'}(1_{HM'}) \circ THg.$$

Hence the diagram

$$
\begin{array}{ccc}
THM & \xrightarrow{\alpha_{HM,M}(1_{HM})} & M \\
THg \downarrow & & \downarrow g \\
THM' & \xrightarrow{\alpha_{HM',M'}(1_{HM'})} & M'
\end{array}
$$

is commutative. Similarly, if $TN' \xrightarrow{v} M'$ and $TN \xrightarrow{u} M$ are morphisms, we have the equations

$$\alpha^{-1}_{N',M'}(v) \circ f = \alpha^{-1}_{N,M'}(v \circ Tf) \tag{A.3}$$

and

$$Hg \circ \alpha^{-1}_{N,M}(u) = \alpha^{-1}_{N,M'}(g \circ u). \tag{A.4}$$

Hence if $M' = TN'$, $M = TN$, and $g = Tf$, we have

$$
\begin{aligned}
\alpha^{-1}_{N',TN'}(1_{TN'}) \circ f &= \alpha^{-1}_{N,TN'}(Tf) \\
&= HTf \circ \alpha^{-1}_{N,TN}(1_{TN}),
\end{aligned}
$$

so the diagram

$$
\begin{array}{ccc}
N & \xrightarrow{\alpha^{-1}_{N,TN}(1_{TN})} & HTN \\
f \downarrow & & \downarrow HTf \\
N' & \xrightarrow{\alpha^{-1}_{N',TN'}(1_{TN'})} & HTN'
\end{array}
$$

is also commutative. In this situation we denote the morphism $\alpha^{-1}_{N,TN}(1_{TN}) : N \to HTN$ by θ_N and the morphism $\alpha_{HM,M}(1_{HM}) : THM \to M$ by μ_M. Then, because of the observations above, $\theta : 1_{\mathcal{D}} \to HT$ and $\mu : TH \to 1_{\mathcal{C}}$ are natural transformations, and then θ is called the *arrow of adjunction* associated with α and μ is its *quasi-inverse*.

Equation (A.1) with $f = \theta_N$ yields

$$
\begin{aligned}
\mu_{TN} \circ T\theta_N &= \alpha_{HTN,TN}(1_{HTN}) \circ T(\theta_N) \\
&= \alpha_{N,TN}(\theta_N) \\
&= \alpha_{N,TN}(\alpha_{N,TN}^{-1}(1_{TN})) \\
&= 1_{TN}.
\end{aligned}
$$

Briefly, we write

$$
\mu T \circ T\theta = 1_T. \tag{A.5}
$$

Similarly, we can apply equation (A.4) to conclude that $H\mu_M \circ \theta_{HM} = 1_{HM}$; that is,

$$
H\mu \circ \theta H = 1_H. \tag{A.6}
$$

Theorem A.2.1. *Given categories $\mathcal{D}$ and $\mathcal{C}$, a pair of covariant functors*

$$
H : \mathcal{C} \rightleftarrows \mathcal{D} : T,
$$

and natural transformations $\theta : 1_{\mathcal{D}} \to HT$ and $\mu : TH \to 1_{\mathcal{C}}$, suppose equations (A.5) and (A.6) are satisfied. Then $\alpha(f) = \mu_M \circ Tf$ for $N \xrightarrow{f} HM$ defines bijections that give H as a right adjoint of T.

Proof. The proof is straightforward, if one notes that the inverse β of α is given by $\beta(g) = Hg \circ \theta_N$ if $TN \xrightarrow{g} M$. ∎

Faithfully Full Adjoints

Theorem A.2.2. *Suppose $H : \mathcal{C} \rightleftarrows \mathcal{D} : T$ is a pair of covariant functors and H is right adjoint to T with arrow of adjunction θ with quasi–inverse μ. Then*

(1) μ is a natural isomorphism if and only if H is full and faithful;
(2) θ is a natural isomorphism if and only if T is full and faithful.

Proof. (1) Assume that μ is a natural isomorphism. If $M \xrightarrow{f_i} M'$, $i = 1, 2$, are morphisms in $\mathcal{C}$ with $Hf_1 = Hf_2$, then

$$
f_1 = \mu_{M'} \circ TH(f_1) \circ \mu_M^{-1} = \mu_{M'} \circ TH(f_2) \circ \mu_M^{-1} = f_2.
$$

Suppose $HM \xrightarrow{g} HM'$. Letting $f = \mu_{M'} \circ T(g) \circ \mu_M^{-1}$, the commutativity of the diagram

$$
\begin{array}{ccc}
HM & \overset{g}{\longrightarrow} & HM' \\
\downarrow \theta_{HM} & & \downarrow \theta_{HM'} \\
HTHM & \overset{HTg}{\longrightarrow} & HTHM' \\
\downarrow H(\mu_M) & & \downarrow H(\mu_{M'}) \\
HM & \overset{Hf}{\longrightarrow} & HM'
\end{array}
$$

(using (A.6)) shows that $H(f) = g$. Hence H is full.

Conversely, fix $M \in \mathcal{C}$. Since H is full and faithful, the composite

$$
\mathrm{Hom}_{\mathcal{C}}(M, M') \overset{H}{\longrightarrow} \mathrm{Hom}_{\mathcal{D}}(HM, HM') \overset{\alpha_{HM,M'}}{\longrightarrow} \mathrm{Hom}_{\mathcal{C}}(THM, M')
$$

for $M' \in \mathcal{C}$ defines a natural isomorphism

$$
\tau : \mathrm{Hom}_{\mathcal{C}}(M, \_) \longrightarrow \mathrm{Hom}_{\mathcal{C}}(THM, \_),
$$

which, according to Corollary A.1.2, is induced by the isomorphism

$$
\tau_M(1_M) = \alpha_{HM,M}(1_{HM}) = \mu_M : THM \to M.
$$

Thus $\mu : TH \to 1_{\mathcal{C}}$ is a natural isomorphism. ∎

The proof of (2) is similar.

A.3. Equivalence of Categories

Definition A.3.1. If $\mathcal{C}$ and $\mathcal{D}$ are categories, a covariant functor $H : \mathcal{C} \longrightarrow \mathcal{D}$ is an equivalence if there is a covariant functor $T : \mathcal{D} \longrightarrow \mathcal{C}$ such that $T \circ H$ and $H \circ T$ are naturally isomorphic to the identity functors $1_{\mathcal{C}}$ and $1_{\mathcal{D}}$, respectively. In this case, T is also an equivalence, and we refer to the equivalence $H : \mathcal{C} \rightleftarrows \mathcal{D} : T$.

Theorem A.3.2. *Let $\mathcal{C}$ and $\mathcal{D}$ be categories and let $T : \mathcal{D} \longrightarrow \mathcal{C}$ be a covariant functor. Then T is an equivalence if an only if T is full and faithful and every object M in $\mathcal{C}$ is isomorphic to an object TN for some object $N \in \mathcal{D}$.*

Proof. Suppose T is an equivalence and let $\eta : 1_{\mathcal{D}} \to HT$ be a natural isomorphism. If $N \overset{f_i}{\longrightarrow} N', i = 1, 2$ are morphisms in $\mathcal{D}$ and we assume that $Tf_1 = Tf_2$, then the commutativity of the diagrams

$$
\begin{array}{ccc}
N & \overset{\eta_N}{\longrightarrow} & HTN \\
f_i \downarrow & & HTf_i \downarrow \\
N' & \overset{\eta_{N'}}{\longrightarrow} & HTN'
\end{array}
$$

and the assumption show that $f_1 = f_2$, so T is faithful. Similarly, H is faithful. If $TN \xrightarrow{g} TN'$ is a morphism, then letting $f = \eta_{N'}^{-1} H g \eta_N$, the commutativity of the diagram

$$
\begin{array}{ccccc}
HTN & \xleftarrow{\eta_N} & N & \xrightarrow{\eta_N} & HTN \\
Hg \downarrow & & f \downarrow & & HTf \downarrow \\
HTN' & \xleftarrow{\eta_{N'}} & N' & \xrightarrow{\eta_{N'}} & HTN'
\end{array}
$$

with isomorphisms η_N, $\eta_{N'}$ shows that $HTf = Hg$ and hence that $Tf = g$ since H is faithful. Hence T is full. Clearly, any object M in $\mathcal{C}$ is isomorphic to the object THM.

Conversely, if M is an object in $\mathcal{C}$ we choose an object HM in $\mathcal{D}$ such that THM is isomorphic to M and we choose an isomorphism $\mu_M : THM \to M$. If $M \xrightarrow{f} M'$ is a morphism in $\mathcal{C}$, then $THM \xrightarrow{\mu_{M'}^{-1} f \mu_M} THM'$ is a morphism in $\mathcal{C}$, and since T is full and faithful, there is a unique morphism $HM \xrightarrow{Hf} HM'$ in $\mathcal{D}$ such that $THf = \mu_{M'}^{-1} f \mu_M$, that is, we have the commutative diagram

$$
\begin{array}{ccc}
THM & \xrightarrow{\mu_M} & M \\
THf \downarrow & & f \downarrow \\
THM' & \xrightarrow{\mu_{M'}} & M'
\end{array}
$$

for exactly one choice of Hf. One verifies that $H : \mathcal{C} \longrightarrow \mathcal{D}$ is a functor and that $\mu : TH \to 1_{\mathcal{C}}$ is a natural isomorphism. ∎

Now, in view of Theorems A.2.2 and A.3.2, we have

Corollary A.3.3. *Suppose $H : \mathcal{C} \rightleftarrows \mathcal{D} : T$ is a pair of covariant functors and H is right adjoint to T with arrow of adjunction $\theta : 1_{\mathcal{D}} \longrightarrow HT$ and its quasi-inverse $\mu : TH \longrightarrow 1_{\mathcal{C}}$. Then $H : \mathcal{C} \rightleftarrows \mathcal{D} : T$ is an equivalence if and only if θ and μ are natural isomorphisms.*

Theorem A.3.4. *Let $\mathcal{C}$ and $\mathcal{D}$ be categories. If $H : \mathcal{C} \rightleftarrows \mathcal{D} : T$ is an equivalence, there exist natural isomorphisms $\theta : 1_{\mathcal{D}} \to HT$ and $\mu : TH \to 1_{\mathcal{C}}$ such that equations (A.5) and (A.6) are satisfied.*

Proof. Suppose $H : \mathcal{C} \rightleftarrows \mathcal{D} : T$ is an equivalence and $\mu : TH \to 1_{\mathcal{C}}$ is a natural isomorphism. Since, for any object N in $\mathcal{D}$,

$$
TN \xrightarrow{\mu_{TN}^{-1}} THTN
$$

is a morphism, and since T is full and faithful, there is a unique morphism $N \xrightarrow{\theta_N} HTN$ such that

$$T\theta_N = \mu_{TN}^{-1}, \tag{A.7}$$

and a simple argument, again since T is full and faithful, shows that θ_N is an isomorphism. If $N \xrightarrow{f} N'$ is a morphism in $\mathcal{D}$, the diagram

$$
\begin{array}{ccc}
TN & \xleftarrow{\mu_{TN}} & THTN \\
Tf \downarrow & & THTf \downarrow \\
TN' & \xleftarrow{\mu_{TN'}} & THTN'
\end{array}
$$

is commutative. Hence, using equation (A.7), we have

$$
\begin{aligned}
Tf &= \mu_{TN'} \circ THTf \circ \mu_{TN}^{-1} \\
&= T\theta_{N'}^{-1} \circ THTf \circ T\theta_N \\
&= T(\theta_{N'}^{-1} HTf\theta_N),
\end{aligned}
$$

so $f = \theta_{N'}^{-1} HTf\theta_N$, that is, the diagram

$$
\begin{array}{ccc}
N & \xrightarrow{\theta_N} & HTN \\
f \downarrow & & HTf \downarrow \\
N' & \xrightarrow{\theta_{N'}} & HTN'
\end{array}
$$

is commutative. Thus $\theta : 1_{\mathcal{D}} \to HT$ is a natural isomorphism. Equation (A.7) shows that equation (A.5) is satisfied. Since μ is a natural isomorphism, the diagram

$$
\begin{array}{ccc}
THM & \xrightarrow{\mu_M} & M \\
TH\mu_M^{-1} \downarrow & & \mu_M^{-1} \downarrow \\
THTHM & \xrightarrow{\mu_{THM}} & THM
\end{array}
$$

is commutative for any object $M \in \mathcal{C}$. Hence we have

$$TH\mu_M = \mu_{THM} = T\theta_{HM}^{-1}$$

where we have used equation (A.7). Since T is full and faithful, we obtain equation (A.6). ∎

Theorem A.3.5. *Let $\mathcal{D}$ and $\mathcal{C}$ be categories and let $T : \mathcal{D} \longrightarrow \mathcal{C}$ be a covariant functor. The following are equivalent.*

(a) T is an equivalence;

(b) T is full and faithful and has a full and faithful right adjoint;

(c) T is full and faithful and has a full and faithful left adjoint.

Proof. $(b) \Rightarrow (a)$ follows from Theorem A.2.2.

$(a) \Rightarrow (b)$ is clear from Theorems A.2.1, A.3.2 and A.3.4.

$(b) \Rightarrow (c)$. Having proved the equivalence of (a) and (b), we note that the symmetry afforded by (a) makes plain the equivalence of (b) and (c).   ∎

Appendix B
Noetherian Serial Rings

A *serial ring* is a semiperfect ring R such that the submodule lattice of each right and each left indecomposable projective R-module is linearly ordered. In this appendix we present results concerning noetherian serial rings that are assumed in some examples given in the text. We let J denote the radical of R.

B.1. Finitely Generated Modules

Lemma B.1.1. *Let R be a noetherian serial ring. If P_R is finitely generated and projective and $K \leq P$, then P has a decomposition into local direct summands*

$$P = P_1 \oplus \cdots \oplus P_n$$

such that

$$K = K \cap P_1 \oplus \cdots \oplus K \cap P_\ell$$

for some $\ell \leq n$.

Proof. The proof is by induction on the composition length $c(K/KJ)$.

First, suppose that e is a primitive idempotent in R and that $K = xeR$. We may assume that $K \subseteq PJ$, since otherwise (see [1, Section 27]) K would be a direct summand of P. Let $N = P/xeR$ so that

$$0 \to xeR \xrightarrow{\subseteq} P \xrightarrow{\eta} N \to 0$$

is a projective cover of N with η the natural map. Now $N = N' \oplus Q''$ where Q'' is projective and N' has no projective direct summands. Letting $Q' \xrightarrow{q} N'$ be a projective cover, we obtain another projective cover

$$(q \oplus 1_{Q''}) : Q' \oplus Q'' \longrightarrow N' \oplus Q'' = N$$

of N. Thus, by uniqueness of projective covers [1, Lemma 17.17] there is an isomorphism

$$f : P \to Q' \oplus Q''$$

such that $\eta = (q \oplus 1_{Q''}) \circ f$. Letting

$$P' = f^{-1}Q' \quad \text{and} \quad P'' = f^{-1}Q'',$$

we see that

$$0 \to xeR \xrightarrow{\subseteq} P' \xrightarrow{\eta|P'} N' \to 0$$

is exact with $P' \xrightarrow{\eta|P'} N'$ a projective cover of N'. But then, since N' has no projective direct summands, the transpose tN' has minimal projective resolution

$$\text{Hom}_R(P', R) \to \text{Hom}_R(eR, R) \to tN' \to 0$$

(see [1, Theorem 32.13]). Since $\text{Hom}_R(eR, R) \cong Re$ is uniserial, we must have $\text{Hom}_R(P', R) \cong Rf$ for some primitive idempotent $f \in R$. Thus $P' \cong \text{Hom}_R(Rf, R) \cong fR$ is uniserial and $K = xeR \subseteq P'$, so the case $c(K/KJ) = 1$ is established, since by the Azumaya-Krull-Schmidt theorem [1, Theorem 12.6] we may renumber the P_i so that

$$P = P' \oplus P_2 \oplus \cdots \oplus P_n.$$

Suppose that $c(K/KJ) = m$, and that the result is true for submodules L of finitely generated projective R-modules with $c(L/LJ) \leq m - 1$. Then clearly $m \leq n$. Let $K = L + xeR \leq P$ for some primitive idempotent $e \in R$, and suppose that

$$P = P_1 \oplus \cdots \oplus P_n$$

with

$$L = L \cap P_1 \oplus \cdots \oplus L \cap P_{m-1}.$$

By the first step in the proof, there is a indecomposable direct summand P' of P with $xeR \leq P'$. Thus, again by the Azumaya-Krull-Schmidt theorem, the given decomposition of P complements P', so that one of the projections $\pi_1, \ldots, \pi_n$ restricts to an isomorphism $(\pi_i|P') : P' \to P_i$ (see [1, Proposition 5.5]). If $i \geq m$, we are done. Otherwise we may assume that $(\pi_1|P') : P' \to P_1$ is an isomorphism. Since P_1 is uniserial, either $\pi_1(xeR) \subseteq L \cap P_1$ or $L \cap P_1 \subseteq \pi_1(xeR)$. In the first case we see that $\pi_1(K) \subseteq K \cap P_1 \subseteq \pi_1(K)$ so that

$$K = K \cap (\pi_1(K) \oplus \pi_2(P) \oplus \cdots \oplus \pi_n(P_n))$$
$$= (K \cap P_1) \oplus (K \cap (\pi_2(P) \oplus \cdots \oplus \pi_n(P_n))),$$

where, modulo its radical, the latter term has length $\leq m - 1$. In the second case, letting $\pi', \pi'_2, \ldots, \pi'_n$ be the projections for the decomposition

$$P = P' \oplus P_2 \oplus \cdots \oplus P_n,$$

the isomorphism $(\pi'|P_1) : P_1 \to P'$ has $\pi'(K) = \pi'(xeR + L \cap P_1) = xeR \subseteq K \cap P' \subseteq \pi'(K)$ so that

$$K = (K \cap P') \oplus (K \cap (\pi'_2(P) \oplus \cdots \oplus \pi'_n(P_n))).$$

In either case the inductive hypothesis applies. ■

Immediately, now, we have

Theorem B.1.2. *If R is a noetherian serial ring, then every finitely generated R-module is a direct sum of local modules.*

Corollary B.1.3. *If R is a noetherian serial ring, then every uniform R module is uniserial.*

Proof. If U_R is uniform and $x, y \in U$, then $xR + yR$ is uniserial by Theorem B.1.2. Thus xR and yR are comparable. ■

Let R be an indecomposable noetherian serial ring, and let $e_1, \ldots, e_n$ be a basic set of primitive idempotents for R. Since R is indecomposable and each $e_i J / e_i J^2$ is simple, the e_i can be renumbered so that $e_i R$ is a projective cover of $e_{i-1}J$ for $i = 2, \ldots, n$ and $e_1 R$ is a projective cover of $e_n J$, if $e_n J \neq 0$. (See [1, Section 32].) When this is the case, $e_1 R, \ldots, e_n R$ is called a *Kupisch series* for R.

We note here that if $e_1 R, \ldots, e_n R$ is a Kupisch series for an indecomposable noetherian serial ring R that is not artinian, then $e_i J \neq 0$ for all $i = 1, \ldots, n$.

Corollary B.1.4. *If R is a noetherian serial ring and e is a primitive idempotent in R, then every non-zero submodule of eR is of the form eJ^k for some $k \in \mathbb{N}$. In particular, $\cap_{k=1}^{\infty} J^k = 0$, and every finitely generated R-module is the direct sum of a projective module and a module of finite length.*

Proof. We may assume that R is indecomposable. Suppose that $\cap_{k=1}^{\infty} eJ^k = N \neq 0$. Then N is an epimorph of some $e_i R$ in a Kupisch series for R, and $e_i R \cong e_j J$. Since $E(eR)$ is uniserial by Corollary B.1.3, $E(eR)/N$ contains a simple, necessarily essential, submodule isomorphic to $e_j R / e_j J$. But this implies $eR / \cap_{k=1}^{\infty} eJ^k$ contains a minimal submodule, which is impossible.

Thus $\cap_{k=1}^{\infty} e J^k = 0$, and if $0 \neq L \leq eR$, then $L = eJ^m$ where m is minimal such that $eJ^m \leq L$. The last statement follows. ∎

Corollary B.1.5. *Every indecomposable noetherian serial ring that is not artinian is hereditary.*

Proof. Let $e_1, \ldots, e_n$ be a basic set of primitive idempotents for R. The preceding discussion shows that since R is not artinian, neither is any $e_i R$. But by Corollary B.1.4, every right ideal I of R is the direct sum of a projective right ideal and an artinian right ideal. In particular, $I_R = P \oplus L$ with P projective and L a right ideal of finite length. But then $L = 0$ and so every right (similarly left) ideal is projective. ∎

It follows that if $e_1 R, \ldots, e_n R$ is a Kupisch series for an indecomposable noetherian serial ring R that is not artinian, then $e_{i-1} J \cong e_1 R$ for $i = 2, \ldots, n$ and $e_n J \cong e_1 R$.

Proposition B.1.6. *Let M_R be a finitely generated (artinian) module over a noetherian serial ring. Then $S = \text{End}(M_R)$ is a semiperfect noetherian (artinian) ring.*

Proof. By Theorem B.1.2 and Corollary B.1.4 M_R is isomorphic to the direct sum of a finite number of modules of the form eR/eI where e is a primitive idempotent and I is an ideal of R. Since $\text{End}((eR/eI)_R) \cong eRe/eIe$ we see that the endomorphism ring of every finitely generated indecomposable module is a semiperfect noetherian (artinian) ring. Hence S is semiperfect by [1, Corollary 27.7]. Similarly, if eR/eI and fR/fK are finitely generated indecomposable with e, f idempotents and I, K ideals of R, then the fRf-eRe bimodule $\text{Hom}_R(eR/eI, fR/fK)$ is isomorphic to a submodule of an epimorphic image of $\text{Hom}_R(eR, fR) \cong fRe$, and so is finitely generated over eRe and over fRf and hence also over the noetherian (artinian) rings $\text{End}((eR/eI)_R) \cong eRe/eIe$ and $\text{End}((fR/fK)_R) \cong fRf/fKf$. It follows that S is a noetherian (artinian) ring. ∎

Proposition B.1.7. *Let M_R, N_R be finitely generated modules over a noetherian serial ring R. Then $\text{Hom}_R(M, N)$ and $\text{Ext}_R^1(M, N)$ are finitely generated both as right $S = \text{End}(M_R)$ modules and as left $S' = \text{End}(N_R)$ modules.*

Proof. By Proposition B.1.6 $\bar{S} = \text{End}(M \oplus N)$ is noetherian. If e, $f \in \bar{S}$ are idempotents, then $e\bar{S}e$ and $f\bar{S}f$ are noetherian, as are $_{e\bar{S}e}\,e\bar{S}f$ and $e\bar{S}f_{f\bar{S}e}$. Thus, with appropriate choices for e and f we see that $\text{Hom}_R(M, N)$ is finitely generated as a right S-module and as a left S'-module. Since R and S are noetherian, taking a projective resolution of M in mod-R yields easily, via the first assertion, that $\text{Ext}^1_R(M, N) \in S'$-mod. To see that $\text{Ext}^1_R(M, N) \in \text{mod-}S$ we may assume that N_R is indecomposable. Then N is uniserial and, since indecomposable injectives are uniserial by Corollary B.1.3, N has an injective resolution,

$$0 \to N_R \to I_0 \to I_1 \to \cdots$$

where each I_j is uniserial. Suppose M_R has finite length. Since each I_j is uniserial, the trace of M in I_j is a finite length submodule K_j of I_j, so $\text{Hom}_R(M, I_j) = \text{Hom}_R(M, K_j)$ is a finitely generated S-module by the first assertion. Since S is noetherian it follows that $\text{Ext}^1_R(M, N) \in \text{mod-}S$. In general, applying Corollary B.1.4, if $M_R = P \oplus M_1$ where P is projective and M_1 has finite length, one obtains easily that $\text{Ext}^1_R(M, N) \in \text{mod-}S$ via $\text{Ext}^1_R(P, N) = 0$ and the preceding argument. ∎

B.2. Injective Modules

Suppose that $e_1 R, \ldots, e_n R$ is a Kupisch series for an indecomposable noetherian serial ring R that is not artinian, and let $S_i = e_i R/e_i J$ for $i = 1, \ldots, n$. Then by Corollary B.1.4 the submodules of $e_i R$ and Re_i must be

$$e_i R > e_i J > e_i J^2 > \cdots \quad \text{and} \quad Re_i > Je_i > J^2 e_i > \cdots .$$

Thus the composition factors of $e_i R$ are, from the top down,

$$S_i, S_{i+1}, \ldots, S_n, S_1, S_2, \ldots, S_n, \ldots .$$

It follows from Corollary B.1.3 that the indecomposable injective R-modules are also uniserial. There are just $n + 1$ of these on each side. In Mod-R they are

$$E_1 = E(S_1), \ldots, E_n = E(S_n) \quad \text{and} \quad E_0$$

with $\text{Soc}(E_0) = 0$. Each E_i with $1 \le i \le n$ is artinian, its submodules are

$$0 < \text{Soc}(E_i) < \text{Soc}^2(E_i) < \cdots ,$$

where $\text{Soc}^k(E_i) = \ell_{E_i}(J^k)$, and the composition factors of E_i are, from the bottom up,

$$S_i, S_{i-1}, \ldots, S_1, S_n, S_{n-1}, \ldots, S_1, \ldots$$

whereas the composition factors of E_0 are, in ascending order,

$$\ldots, S_n, S_{n-1}, \ldots, S_1, S_n, S_{n-1}, \ldots, S_1, \ldots.$$

From these observations we glean

Proposition B.2.1. *Let R be an indecomposable noetherian serial ring that is not artinian. Then every proper factor of an indecomposable injective R-module is the injective envelope of its socle, and R has a unique indecomposable non-artinian injective right (or left) module E_0, all of whose proper submodules are indecomposable projective modules. Moreover, every indecomposable injective right (or left) R-module is isomorphic to a factor of E_0.*

Corollary B.2.2. *Let R be an indecomposable noetherian serial ring that is not artinian, and suppose that R has n isomorphism classes of simple right modules and indecomposable injective right modules $E_1, \ldots, E_n, E_0$ with all but E_0 artinian. If $i \in \{1, \ldots, n\}$, then for each set A there are sets B, C such that $E_i^A \cong E_i^{(B)} \oplus E_0^{(C)}$.*

Proof. Since R is semiperfect and J is finitely generated, it follows that

$$\text{Soc}\left(E_i^A\right) = \ell_{E_i^A}(J) = \ell_{E_i}(J)^A = \text{Soc}(E_i)^A.$$

Thus since $\text{Soc}(E_i)e_j = 0$ if $j \neq i$, we see that $\text{Soc}(E_i^A) \cong S_i^{(B)}$ for some set B. But then (see [1, Proposition 18.13]) $E_i^A \cong E_i^{(B)} \oplus E$ where $\text{Soc}(E) = 0$, so (see [1, Theorem 25.6]) there must be a set C with $E \cong E_0^{(C)}$. ∎

Proposition B.2.3. *If R is a left (equivalently, right) linearly compact indecomposable non-artinian noetherian serial ring, then R has self-duality.*

Proof. Assume that R is self-basic and left linearly compact, and let $E = E_1 \oplus \cdots \oplus E_n$ be the minimal left cogenerator. Then E is artinian and hence linearly compact ([81, Corollary 3.2]). Thus, letting $S = \text{End}(_R E)$, the bimodule $_R E_S$ induces a Morita duality and S_S is linearly compact (see Section 4.4). Now, letting $A_k = \mathbf{r}_E(J^k)$, it is apparent that A_k is the minimal cogenerator over R/J^k and that the bimodule $_{R/J^k} A_{kS/\mathbf{r}_S(A_k)}$ induces a Morita duality. But R/J^k is a basic QF-ring (see [1, Section 32.6]), so we must have $_{R/J^k} A_k \cong {}_{R/J^k} R/J^k$. Thus, as rings

$$S/\mathbf{r}_S(A_k) \cong \text{End}(_{R/J^k} A_k) \cong R/J^k.$$

Now both $\{J^k \mid k \geq 1\}$ and $\{\mathbf{r}_S(A_k) \mid k \geq 1\}$ are downward directed sets of ideals with, by Corollary B.1.4, $\cap_k J^k = 0$, and hence $\cap_k \mathbf{r}_S(A_k) = 0$. Therefore, since $_R R$ and S_S are linearly compact,

$$R \cong \varprojlim R / J^k \cong \varprojlim S / \mathbf{r}_S(A_k) \cong S. \qquad \blacksquare$$

The structure of noetherian serial rings, as well as several of the results in this appendix, were given by Warfield in [79]. In particular it is proved there that every indecomposable noetherian serial ring that is not artinian is Morita equivalent to an $n \times n$ $[D : \text{M}]$ upper triangular matrix ring over a local noetherian uniserial ring D with maximal ideal M, that is, a full ring of $n \times n$ matrices over D whose entries below the main diagonal belong to M. Moreover, it follows from Proposition B.2.3 and [81, Theorem 4.3, Lemma 4.9 and Proposition 3.3] that $[D : \text{M}]$ has self-duality if and only if D is linearly compact.

Bibliography

[1] F. W. Anderson and K. R. Fuller. *Rings and Categories of Modules*. Springer–Verlag, Inc., New York, Heidelberg, Berlin, second edition, 1992.

[2] L. Angeleri Hügel. Finitely cotilting modules. *Comm. Algebra*, 28:2147–2172, 2000.

[3] L. Angeleri Hügel, A. Tonolo, and J. Trlifaj. Tilting preenvelopes and cotilting precovers. *Algebras and Representation Theory*, 4:155–201, 2002.

[4] I. Assem. Torsion theories induced by tilting modules. *Can. J. Math.*, 36:899–913, 1984.

[5] I. Assem. Tilting theory – an introduction. *Topics in Algebra, Banach Center Pub.*, 26:127–180, 1990.

[6] G. Azumaya. A duality theory for injective modules. *Amer. J. Math.*, 81:249–278, 1959.

[7] S. Bazzoni. Cotilting modules are pure injective. *Proc. Amer. Math. Soc.*, 31:3665–3672, 2003.

[8] K. Bongartz. Tilted algebras. *Springer–Verlag LNM*, 903:26–38, 1981.

[9] S. Brenner and M. C. R. Butler. Generalizations of the Bernstein–Gelfand–Ponomarev reflection functors. *Springer–Verlag LNM*, 832:103–170, 1980.

[10] W. D. Burgess and X. Du. On the torsion parts of a torsion theory counter equivalence. (preprint).

[11] H. Cartan and S. Eilenburg. *Homological Algebra*. Princeton University Press, Princeton, New Jersey, 1956.

[12] S. U. Chase. Direct products of modules. *Trans. Amer. Math. Soc.*, 97:457–473, 1960.

[13] P. M. Cohn. *Morita equivalence and duality*. Mathematical Notes, Queen Mary College, Univ. of London, London, 1966.

[14] R. R. Colby. A generalization of Morita duality and the tilting theorem. *Comm. Alg.*, 17(7):1709–1722, 1989.

[15] R. R. Colby. A cotilting theorem for rings. In *Methods in Module Theory*, pages 33–37. M. Dekker, New York, 1993.

[16] R. R. Colby, R. Colpi, and K. R. Fuller. A note on cotilting modules and generalized Morita duality. *Venizia 2002 Conference Proceedings*. (to appear).

[17] R. R. Colby and K. R. Fuller. QF–3′ rings and Morita duality. *Tsukuba J. Math.*, 8:183–188, 1984.

[18] R. R. Colby and K. R. Fuller. Tilting, cotilting, and serially tilted rings. *Comm. Algebra*, 18:1585–1615, 1990.

[19] R. R. Colby and K. R. Fuller. Tilting and torsion theory counter equivalences. *Comm. Algebra*, 23(13):4833–4849, 1995.

[20] R. R. Colby and K. R. Fuller. Hereditary torsion theory counter equivalences. *J. Algebra*, 183:217–230, 1996.

[21] R. R. Colby and K. R. Fuller. Costar modules. *J. Algebra*, 242:146–159, 2001.

[22] R. R. Colby and K. R. Fuller. Weak Morita duality. *Comm. Algebra*, 31:1859–1879, 2003.

[23] R. Colpi. Dualities induced by cotilting modules. *Venizia 2002 Conference Proceedings*. (to appear).

[24] R. Colpi. Some remarks on equivalences between categories of modules. *Comm. Algebra*, 18:1935–1951, 1990.

[25] R. Colpi. Tilting modules and ∗–modules. *Comm. Algebra*, 21(4):1095–1102, 1993.

[26] R. Colpi. Cotilting bimodules and their dualities. Interactions between ring theory and representations of algebras (Murcia). In *Lecture Notes in Pure and Appl. Math.*, volume 210, pages 81–93. Marcel Dekker, New York, 2000.

[27] R. Colpi and G. D'Este. Equivalences represented by faithful non-tilting ∗–modules. *Canadian Math. Soc., Conference Proc.*, 24:103–110, 1998.

[28] R. Colpi, G. D'Este, and A. Tonolo. Quasi–tilting modules and counter equivalences. *J. Algebra*, 191:461–494, 1997.

[29] R. Colpi and K. R. Fuller. Cotilting modules and bimodules. *Pacific J. Math.*, 192:275–291, 2000.

[30] R. Colpi and C. Menini. On the structure of ∗–modules. *J. Algebra*, 158:400–419, 1993.

[31] R. Colpi, A. Tonolo, and J. Trlifaj. Partial cotilting modules and the lattices induced by them. *Comm. Algebra*, 25:3225–3237, 1997.

[32] R. Colpi and J. Trlifaj. Tilting modules and tilting torsion theories. *J. Algebra*, 178:614–634, 1995.

[33] G. D'Este. Reflexive modules are not closed under submodules. In *Lecture Notes in Pure and Appl. Math.*, volume 224, pages 53–64. Marcel Dekker, New York, 2002.

[34] G. D'Este and D. Happel. Representable equivalences are represented by tilting modules. *Rend. Sem. Mat. Univ. Padova*, 83:77–80, 1990.

[35] P. C. Eklof and A. H. Mekler. *Almost Free Modules (Set Theoretic Methods)*. North-Holland, Amsterdam, New York, Oxford, Tokyo, 1990.

[36] L. Fuchs. *Infinite Abelian Groups*, volume II. Academic Press, Inc., New York, London, 1973.

[37] L. Fuchs and L. Salce. *Modules over Non–Noetherian Domains*. American Mathematical Society, Providence, 2001.

[38] K. Fuller. Density and equivalence. *J. Algebra*, 29:528–550, 1974.

[39] K. R. Fuller. A note on quasi–duality. In *Abelian Groups, Module Theory, and Topology*, volume 201 of *Lect. Notes Pure Appl. Math.* Marcel Dekker, New York, 1998.

[40] K. R. Fuller and W. Xue. On quasi–duality modules. *Comm. Algebra*, 28:1919–1937, 2000.

[41] J. L. Gómez Pardo. Counterinjective modules and duality. *J. Pure and Appl. Algebra*, 61:165–179, 1989.

[42] J. L. Gómez Pardo, P. A. Guil Asensio, and R. Wisbauer. Morita dualities induced by the M-dual functors. *Comm. Algebra*, 22:5903–5934, 1994.

[43] P. Griffith. On a subfunctor of Ext. *Arch. Math.*, XXI:17–22, 1970.

[44] D. Happel. *Triangulated Categories in the Representation Theory of Finite Dimensional Algebras*, volume 118. London Math. Soc. Lecture Notes Series, 1988.

[45] D. Happel, I. Reiten, and S. Smalø. *Tilting in Abelian Categories and Quasi-tilted Algebras*, volume 120. Mem. Amer. Math. Soc., Providence, 1996.

[46] D. Happel and C. M. Ringel. Construction of tilted algebras. *Springer–Verlag LNM*, 903:125–144, 1981.

[47] D. Happel and C. M. Ringel. Tilted algebras. *Trans. Amer. Math. Soc.*, 274:399–443, 1982.

[48] M. Hoshino. Tilting modules and torsion theories. *Bull. London Math. Soc.*, 14:334–336, 1982.

[49] M. Hoshino. Splitting torsion theories induced by tilting modules. *Comm. Algebra*, 11:427–441, 1983.

[50] N. Jacobson. *Lectures in Abstract Algebra*, volume 2. Van Nostrand, New York, 1953.

[51] J. P. Jans. Duality in Noetherian rings. *Proc. Amer. Math. Soc.*, 12:829–835, 1961.

[52] C. U. Jensen and H. Lenzing. *Model Theoretic Algebra*, volume 2. Gordon and Breach Science Publishers, New York, 1989.

[53] H. Krause and M. Saorín. On minimal approximations of modules. *Contemp. Math.*, 299:227–236, 1998.

[54] H. Lenzing. Endlich präsentierbare Moduln. *Arch. Math.*, 20:262–266, 1969.

[55] R. N. S. MacDonald. Representable dualities between finitely closed subcategories of modules. *Can. J. Math.*, 31:465–475, 1979.

[56] S. MacLane. *Homology*. Academic Press, Inc., Publishers, New York, 1963.

[57] S. Maclane. *Categories for the Working Mathematician*. Graduate Texts in Mathematics. Springer–Verlag, New York Heidelberg Berlin, 1971.

[58] F. Mantese. Generalizing cotilting dualities. *J. Algebra*, 236:630–644, 2001.

[59] F. Mantese, P. Ruzicka, and A. Tonolo. Cotilting versus pure-injective modules. *Pacific J. Math.* (to appear).

[60] E. Matlis. Reflexive domains. *J. Algebra*, 8:1–33, 1967.

[61] C. Menini and A. Orsatti. Good dualities and strongly quasi-injective modules. *Ann. Mate. Pure Appl. (IV)*, 127:187–230, 1981.

[62] C. Menini and A. Orsatti. Representable equivalences between categories of modules and applications. *Rend. Sem. Mat. Univ. Padova*, 82:203–231, 1989.

[63] B. Mitchell. *Theory of Categories*. Pure and Applied Mathematics. Academic Press, Inc., New York and London, 1965.

[64] Y. Miyashita. Tilting modules of finite projective dimension. *Math. Z.*, 193:113–146, 1986.

[65] K. Morita. Duality for modules and its applications to the theory of rings with minimum condition. *Sci. Rep. Tokyo Kyoiku Daigaku Ser. A*, 6:83–142, 1958.

[66] B. J. Müller. Linear compactness and Morita duality. *J. Algebra*, 16:60–66, 1970.

[67] N. Popescu. *Abelian Categories with Applications to Rings and Modules*. Academic Press, Inc., New York, 1973.

[68] G. Puninski. *Serial Rings*. Kluwer Academic Publishers, Dordrecht, The Netherlands, 2001.

[69] J. J. Rotman. *An Introduction to Homological Algebra*. Academic Press, Inc., New York, San Francisco, London, 1979.

[70] F. L. Sandomierski. Linearly compact modules and local Morita duality. In R. Gordon, editor, *Ring Theory*, pages 333–346. Academic Press, Inc., New York, 1972.

[71] S. O. Smalø. Torsion theories and tilting modules. *Bull. London Math. Soc.*, 16:518–522, 1984.

[72] B. Stenstrom. *Rings of Quotients. Springer–Verlag LNM* vol. 76, New York, Heidelberg, Berlin, 1970.

[73] R. G. Swan. *Algebraic K–Theory*, volume 76. *Springer–Verlag LNM* vol. 76, Berlin, Heidelberg, New York, 1968.

[74] H. Tachikawa. Representations of trivial extensions of hereditary algebras. *Springer–Verlag LNM*, 832:579–599, 1980.

[75] A. Tonolo. Generalizing Morita duality: A homological approach. *J. Algebra*, 232:282–298, 2000.

[76] A. Tonolo. On a finitistic cotilting–type duality. *Comm. Algebra*, 30:5091–5106, 2002.

[77] J. Trlifaj. On *–modules generating the injectives. *Rend. Sem. Mat. Univ. Padova*, 88:211–220, 1992.

[78] J. Trlifaj. Every *–module is finitely generated. *J. Algebra*, 169:392–398, 1994.

[79] R. B. Warfield. Serial rings and finitely presented modules. *J. Algebra*, 37:187–222, 1975.

[80] R. Wisbauer. *Foundations of Module and Ring Theory*. Gordon and Breach Science Publishers, Philadelphia, 1991.

[81] W. Xue. *Rings with Morita Duality. Springer–Verlag LNM* vol. 1523, Berlin, Heidelberg, New York, 1992.

[82] K. Yamagata. Extensions over hereditary artinian rings with self-dualities, I. *J. Algebra*, 73:386–433, 1981.

[83] J. M. Zelmanowitz and W. Jansen. Duality modules and Morita duality. *J. Algebra*, 129:257–277, 1989.

Index

∗-module, 18
∗-module
  weak, 15
$[D :\text{M}]$ upper triangular matrix ring, 56

abelian subcategory, 86
$\text{Add}(V)$, 8
adjoint functor
  arrow of adjunction, 12
  quasi-inverse, 12
adjoint functor, 132
  arrow of adjunction, 133
  quasi-inverse, 133
almost disjoint, 95
artin algebra, 9, 33

balanced
  faithfully, 22
bimodule
  tilting, 33

$c_i(M)$, 52
Cogen, 10
cogen, 10
Copres, 14
copresented, 14
  finitely copresented, 68
copresented
  semi-finitely copresented, 68
costar module, 69
cotilting bimodule, 97
cotilting module, 92
  finitistic, 115
cotilting theorem, 113

Δ, 66, 92
dense, 26
duality
  artin algebra duality, 33
duality, 66
  representable duality, 66
duality module, 84

equivalence, 12, 13, 15, 17–19, 22, 24, 25, 135
  $\approx$, 12
  induced by a progenerator, 25
  representable, 12

Γ, 86, 92
Gen, 2
gen, 2
generalized Morita duality, 109
GMD, 109
Grothendieck group
  on mod-R, 49
  on proj-R, 54

injective relative to, 75

Kupisch series, 141

linearly compact, 78

Mod, 1
mod, 1
module
  cotilting module, 92
  quasi-duality module, 75
Morita duality, 84, 103
  generalized, 109
  weak, 109
Morita equivalence, 22, 24

partial tilting module, 36
  generalized, 37
$\perp$
  $^{\perp}U$, 91
$\perp$
  $V^{\perp}$, 1
Pres, 14
presented, 14
$\text{Prod}(V)$, 8
progenerator, 22
proj.dim., 1
projective dimension, 1

151